GLOBAL CHANGE
IN THE
MOUNTAINS

Martin Price

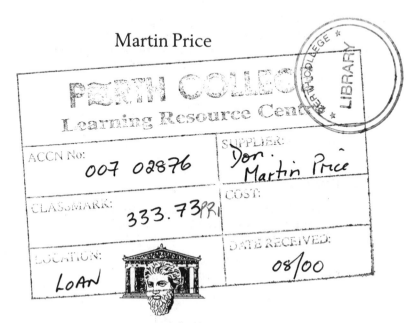

The Parthenon Publishing Group
International Publishers in Medicine, Science & Technology

NEW YORK   LONDON

Published in the USA by
The Parthenon Publishing Group Inc.
One Blue Hill Plaza
PO Box 1564, Pearl River
New York 10965, USA

Published in the UK and Europe by
The Parthenon Publishing Group Limited
Casterton Hall, Carnforth
Lancs., LA6 2LA, UK

Library of Congress Cataloging-in-Publication Data
European Conference on Environmental and Societal Change in Mountain
    Regions (1997 : Oxford, England)
      Global change in the mountains : proceedings of the European
    Conference on Environmental and Societal Change in Mountain
    Regions : Oxford, UK, 18–20 December 1997 / edited by Martin F.
    Price, Thomas H. Mather, and Elliot C. Robertson.
      p.   cm.
    Includes bibliographical references and index.
    ISBN 1-85070-062-1
      1. Mountains—Environmental aspects—Congresses. 2. Climatic
    changes—Environmental aspects—Congresses. I. Price, Martin F.
    II. Mather, Thomas H. III. Robertson, Elliot C. IV. Title.
    05501.2.E87 1997
    333.73—dc21   98-48934
                                                  CIP

British Library Cataloguing in Publication Data
Global change in the mountains : proceedings of the European Conference on
    Environmental and Societal Change in Mountain Regions, Oxford, UK
    18–20 December 1997
    1. Mountain ecology – Congresses 2. Global environmental change –
    Congresses 3. Mountains – Social aspects – Congresses
    I. Price, Martin F. II. Mather, Thomas H. III. Robertson, Elliot C.
    IV. European Conference on Environmental and Societal Change in
    Mountain Regions (1997 : Oxford, England)
    333.7′84

ISBN 1-85070-062-1

Typeset by Martin Lister Publishing Services, Carnforth, UK
Printed and bound by J.W. Arrowsmith Ltd., Bristol, UK

GLOBAL CHANGE
IN THE
MOUNTAINS

This volume is dedicated to the memory of Jennifer Nagy (née McConnell) and Graham E. Clarke, two scientists who devoted their lives to mountain people and environments. Both participated in the Oxford conference, but died tragically in early 1998.

Pierre Mathy and Steven Morgan
ENRICH Secretariat, Directorate-General XII-D
European Commission, 1049 Brussels
Belgium

Foreword

These proceedings are a major product of a project which is one of 19 undertaken as a result of a first call for proposals in October 1996 under the European Union's (EU) European Network for Research in Global Change, known as ENRICH. ENRICH is co-sponsored and funded by three Specific Programmes under the EU Fourth Framework Programme in Research and Technological Development (RTD): Environment and Climate, Marine Science and Technology, and Cooperation with Third Countries and International Organisations.

The focus of ENRICH is on global change, and fundamental to its mission is the facilitation of networking between European researchers and their colleagues in the wider international research community, and also in promoting links between researchers and end users of such research. Since 1996, ENRICH has sought to do this through galvanising the scientific community through calls for proposals. This bottom-up approach is instrumental in providing a transparent and fair mechanism for the best European and international scientists to bid for scarce research networking and coordination funds. This 'Mountain Changes' project has been successful in fulfilling both sets of demanding requirements.

These proceedings result from a major international conference which focused on the role and significance of global change for mountain regions. A total of 128 participants from 30 countries on four continents took part in the Conference, held at the University of Oxford in December 1997. The quality of both the presentations and the poster sessions convinced the organisers of the need to publish this material as a fitting permanent record of the Conference. In addition, the Organising Committee produced a booklet summarising the key conclusions and recommendations which came from these papers and the three-day Conference. The text of this booklet is also reproduced at the beginning of these Proceedings.

As both documents illustrate very clearly, mountain areas are of global significance, both in terms of their role in the Earth system vis-à-vis the climate, functioning of ecosystems, and transport of materials and water, but also in terms of their role in providing homes and jobs for around 10% of the Earth's population. The integration of earth system and human system perspectives underpins the philosophy of the Fifth EU RTD Framework Programme. In this regard, this project has hopefully made a useful contribution to the development of an international research agenda on an important component of the

v

Global Change, Climate and Biodiversity Key Action in the Fifth Framework Programme.

Whilst ENRICH has provided the core funding for this project, important support was given by other programmes and organisations, both through direct sponsorship to the organisers or through funding of individual travel and living expenses of individual Conference participants. ENRICH recognises and encourages this co-funding and partnership approach, and would like to thank all the co-sponsors, large and small, from all over the world, for their contributions.

Contents

Alps and Northern Europe

Contents

Contents

Martin F. Price
Mountain Regions Programme, Environmental Change Unit
University of Oxford, 11 Bevington Road, Oxford OX2 6NB
United Kingdom

Preface and Acknowledgements

This volume is the product of a major international and interdisciplinary effort which took place over six months at the end of 1997 and the beginning of 1998. As Pierre Mathy and Steven Morgan outline in their Foreword, this initiative was primarily supported by the European Union's ENRICH programme, itself an important example of collaboration within the European Commission. Without this support, which was approved in record time, it is unlikely that the European Conference on Environmental and Societal Change in Mountain Regions would have taken place. It certainly would have been a very different event, and I would like to reiterate my thanks to ENRICH for their vital assistance.

One of the key elements of the funding provided by ENRICH was for meetings of a Programme Advisory Committee to design the conference and to synthesise its conclusions and recommendations. The Programme Advisory Committee met for the first time under my chairmanship at the Potsdam Institut für Klimafolgenforschung, Germany, on 10 and 11 October 1997. The committee was interdisciplinary and international, including members from Austria, the Czech Republic, France, Germany, India, Norway, Spain, and the UK. All of its members are closely involved in mountain components of European and international research and/or development programmes and/or have expertise in the themes of the conference. Although most of the members of the committee had never met previously, they rapidly coalesced into an efficient team who prepared a well-structured programme and took on the various responsibilities for ensuring that the conference and its follow-up met our desired objectives.

Two of the committee members represented partner organisations which provided financial assistance and other resources for the meetings of the committee: Alfred Becker (Potsdam Institut für Klimafolgenforschung, Germany) and François Gillet (Pôle Grenoblois d'Etudes et de Recherche pour la Prevention des Risques Naturels, Grenoble, France). I would like to thank them and their organisations, and also the other members of the committee: Inger-Marie Bjønness (Ministry of Foreign Affairs, Oslo, Norway); David Collins (School of Geography, University of Oxford, UK); Jordi Corominas (Departament d'Enginyeria del Terreny, ETS Enginyers de Camins, Barcelona, Spain); Rita Gardner (Royal Geographical Society, London, UK); Georg Grabherr (Institut für Pflanzenphysiologie, Universität Wien, Austria); Bill Heal (Institute of Ecology and Resource Management, University of Edinburgh, UK);

Jan Kalvoda (Katedra fyzické geografie a geoekologie, University Karlovy, Prague, Czech Republic); P.S. Ramakrishnan (School of Environmental Sciences, Jawaharlal Nehru University, Delhi, India); Matthias Winiger (Geographisches Institut, Universität Bonn, Germany); and Friedrich Zimmermann (Geographisches Institut, Universität Graz, Austria). I would also like to thank the two alternate members of the committee who replaced members who could not attend meetings: Bernard Debarbieux (Institut de Géographie Alpine, Grenoble, France), and Des Thompson (Scottish Natural Heritage, Edinburgh, UK).

The Conference itself took place at Keble College and the School of Geography, University of Oxford, on 18–20 December 1998. In spite of the short time available for planning and organisation, 128 people from 30 countries on four continents participated. This broad participation was made possible not only by the ENRICH funding, but also by support from the Arkleton Trust, the ARTERI Concerted Action, the International Geographical Union's Commission on Mountain Geoecology and Sustainable Development, and The British Council. I would like to recognise this support, together with financial, logistic, and other resources provided by the Mount Everest Foundation, the Natural Environment Research Council, the Rhodes Trust, and the University of Oxford's Environmental Change Unit and School of Geography. The conference was a very stimulating event, and rather successful in achieving its objectives, which are presented in the following section.

One of the benefits of the ENRICH funding for the Programme Advisory Committee was that we did not have to work into the wee hours after the conference banquet to prepare final syntheses and recommendations. The Conference concluded with summaries of the findings of the regional and thematic workshops. Over the next two months, these were reviewed and synthesised by members of the Committee, which met for the last time at the Institut de Géographie Alpine, Grenoble, France, on 20 and 21 February 1998. The main purposes of the meeting were to produce a short document of conclusions and recommendations which could be rapidly and widely disseminated, and to make final decisions regarding other publications. The Conclusions and Recommendations were published and distributed in March 1998, and also posted on the Internet. They are also reproduced as the first part of these proceedings, following this Preface. In addition to these proceedings, a number of conference participants have been invited to submit full-length papers, to be considered for inclusion in special issues of refereed journals to be edited by Committee members. At the time of writing, these arrangements have not been finalised.

Following the Conclusions and Recommendations, these proceedings are divided into nine sections, each containing extended abstracts of presentations and posters at the conference. The first section includes a series of global perspectives: policy-oriented, research-oriented, and epistemological. All underline the importance of jointly considering the human and natural aspects of mountain systems. The second section presents a number of papers on programmes and methods at various scales. Each of the remaining sections comprises a set of papers describing research in a different region of the world. These regions were defined by the Programme Advisory Committee according to a number of geographical and cultural criteria, and proved quite coherent as the basis for discussions in workshops during the conference, as noted in the introductions to each section. One of the important lessons of these workshops was that people working in a particular region, no matter how diverse their disciplines, found much in common – much more, in fact, than in the thematic workshops which attempted to bring together people from similar disciplines, but working in different mountain regions. This is clearly something to be considered when developing future programmes on global change in the mountains.

Overall, these proceedings show the great diversity of research taking place in mountain regions at the end of the twentieth century. They are clearly a snapshot, reflecting the interest and ability of individuals to participate in the conference, with a bias towards perspectives from Europe (as predetermined by the title of the conference). Nevertheless, they show the need for collaboration between scientists from various disciplines to understand the complex issues which characterise mountain environments and those who live in, depend, and enjoy them. As noted in the Conclusions and Recommendations, collaboration requires well-structured frameworks, of which one vital element is accessible information. The ENRICH support was a vital catalyst, and is also supporting the development of a database of ongoing research in mountain regions. I hope that all those who read this preface and are undertaking such research will contribute to this database, accessible via Internet at http://www.ecu.ox.ac.uk/ Mountains/database/ hostpage.htm. Since the conference, a new framework for collaboration has been established: a Mountain Research Group under the auspices of the Royal Geographical Society with the Institute of British Geographers. I hope that this will provide a new focal point for those who work in mountain regions to contact each other and find new ways of working together.

To conclude, I would like to thank, first, David Bloomer and Helen Lee of Parthenon Publishing for their support to the conference and for publishing these proceedings and, second, my two co-editors. The proceedings could not have been produced so quickly and efficiently without Tom Mather's assistance. And, as was clear to all those involved in the conference and its follow-up, it would not have been the success it was without Elliot Robertson's dedication. Mountain people have a tradition of cooperation - and this has been equally evident in this venture involving a very diverse group of people who seek to understand mountain people and their environments in this period of change.

Conclusions and Recommendations

1. Mountain regions as a focus for global change research

Mountain regions cover about one-fifth of the continents and greatly influence regional and continental atmospheric circulation as well as water and energy cycles. They provide the direct life-support base for about a tenth of humankind, and indirectly affect the lives of more than half.[1] Such indirect effects refer to the provision of goods and services including desired volumes of high-quality water, hydro-electric power, agricultural and forest products, and landscapes for tourism and recreation.

Globally important mountain resources[2]

- **Water**: the world's largest rivers originate in the mountains, which provide water to at least half of the world's population, both within the mountains and downstream. While mountain rivers are responsible for erosion and sedimentation problems, they also transport much beneficial sediment. They are also major sources of hydro-electric power. Many cross national boundaries.

- **Biodiversity**: mountains are core areas of global biodiversity, especially in tropical and subtropical zones. Many of the world's crops originate in mountains, which are also of crucial importance as refuges for species, e.g., relicts, endemics, or eradicated from the lowlands.

- **Landscapes for recreation and tourism**: recreation and tourism is the world's largest industry, and much of this centres on mountains. Over half of the world's population lives in urban areas, and this proportion will continue to grow, leading to increasing demands for recreation and tourism.

- **Mountain societies**: the cultures, identities, local knowledge, technical expertise, and traditions of mountain communities are of considerable benefit to the global community. They are under significant threat through the forces of globalisation.

The 1995 Assessment of the Inter-governmental Panel on Climate Change (IPCC) devoted a chapter to mountain regions, noting their major significance, and the important consequences for ecological and social systems likely to be associated with climate change, particularly through changes in the frequency of extreme events; e.g., in relation to water resources, biodiversity, agriculture, forestry, and tourism.[3] Such impacts may include:

---

[1] Ives, J.D. (1992) Preface P.B. Stone (ed.) The State of the World's Mountains, Zed Books, London: xiii-xvi.

[2] Messerli, B. (1997) The global mountain problematique. Abstracts, European Conference on Environmental and Societal Change in Mountain Regions: 2–3.

[3] Beniston, M. and Fox, D.G. (eds.) (1996) Impacts of climate change on mountain regions Watson, R.T. et al. (eds.) Climate Change 1995. Impacts, Adaptations and Mitigation of Climate Change: Scientific-Technical Analyses. Cambridge University Press, Cambridge: 191–213.

- increased frequencies of natural hazards, with associated losses of life and property;
- changes in precipitation, snow- and glacier accumulation and melt regimes, affecting the quality and quantity of water supplies and the generation of hydro-electric power;
- shifts in ecological and agro-ecological zones, significantly affecting the survival of semi-natural ecosystems and threatened/endangered species, and influencing possibilities for the cultivation of crops and trees;
- changes in resources essential for tourism, such as attractive landscapes, snow for skiing, and desirable weather conditions.

Mountains: steep slopes and sharp gradients

Although the world's mountains differ considerably in many characteristics, all have steep slopes and significant relief (topographic variation). These give rise to some of the sharpest gradients found on Earth. Related characteristics include:[4]

- rapid and systematic changes in climatic variables (especially temperature and precipitation) over very short distances;
- sharp climatic changes independent of photoperiod (length of daylight) and often soil type;
- greatly enhanced direct runoff and erosion, strongly influencing overall hydrology;
- systematic variation of other environmental properties (e.g., soil depth and structure, CO_2, UV-B) with elevation.

With their steep topographical gradients, mountains present unique challenges for global change research, and are:

- particularly valuable in providing basic understanding of hydrological and ecological responses to global change;
- particularly susceptible to the impacts of a rapidly changing climate, often coupled with land-use pressures;
- critical locations for the detection and study of the signals of climate change and its impacts on hydrological, ecological, and societal systems.

For at least the next few decades, social, economic, and political factors are likely to be at least as important as environmental changes as driving forces of global change in mountain areas. These linked factors include a number of processes, including:

- the incorporation of mountain economies into wider economies, up to the global scale;
- seasonal and permanent migration, particularly from, but also into, mountain areas; and the associated flows of financial and other resources;
- the increasing influence of urban processes and perspectives on mountain regions, both through urbanisation, but also as a result of new communications (including transport and various electronic media);

---

[4] Becker, A. and Bugmann, H. (eds.) (1997) Predicting Global Change Impacts on Mountain Hydrology and Ecology: Integrated Catchment Hydrology/Altitudinal Gradient Studies. IGBP Report 43, IGBP Secretariat, Stockholm.

- international legal instruments and other initiatives, and national government policies which lead to changes in the location of decision-making; in some cases towards the global, in some cases towards the local (cf. decentralisation, subsidiarity).

Research on global change in mountain regions should address all of the diverse driving forces, both environmental and societal. In Europe, this is of particular importance in the context of the expansion of the European Union (EU) and the potential restructuring of the Common Agricultural Policy, as well as other instruments within EU environmental and cohesion policies, for instance towards subsidies and financial support for environmental goals and the economic development of 'less-favoured areas'. These have been major determinants of land-use in mountain areas.

2. Mountains in science and policy: a focus for action

In June 1992, at the UN Conference on Environment and Development (UNCED), held in Rio de Janeiro, the heads of state or government of most of the world's countries endorsed "Agenda 21": a plan for action into the 21st century. Chapter 13 of "Agenda 21" is entitled "Managing Fragile Ecosystems: Sustainable Mountain Development". The inclusion of this chapter meant that, for the first time, mountain regions were accorded equal priority in the global environment-development agenda with other global change topics such as climate change, desertification, or deforestation.

New attention to mountain issues since 1992

- Many international organisations, including United Nations agencies, the World Bank, the Inter-governmental Panel on Climate Change (IPCC), the Consultative Group on International Agricultural Research (CGIAR), Diversitas (involving ICSU, IUBS, SCOPE, UNESCO, IGBP/GCTE and IUMS[5]), and the International Union of Forestry Research Organisations (IUFRO), have increased their attention to mountain issues.
- The International Geosphere-Biosphere Programme (IGBP) has begun work towards a inter-core project on 'Global Change Impacts on Mountain Hydrology and Ecology', associating four core projects and the Global System for Research, Analysis and Training (START); and preliminary work has also begun towards a project on the dynamics of land-use and cover change in the Hindu Kush-Himalaya, within the scope of the Land-use and Land Cover core project of the IGBP and the International Human Dimensions Programme (IHDP).
- Regional institutions concentrating on mountain regions have increased their activities in recent years. These include the International Centre for Integrated Mountain Development (ICIMOD, Nepal), the Consortium for the Sustainable Development of the Andean Ecoregion (CONDESAN), and the institutions associated with the Alpine Convention.

Since 1992, mountain regions have continued to move up both political and scientific agendas. Under the aegis of the Food and Agriculture Organization of

---

[5] See list of acronyms.

the United Nations (FAO), regional inter-governmental consultations on sustainable mountain development have been held around the world. The European consultation, held in 1996, involved representatives of 33 countries and the European Union, who endorsed a number of recommendations, including a number on research, training, monitoring, evaluation, and communication.[6] Many countries, both in Europe and on other continents, are implementing policies aiming at sustainable mountain development, and a draft European Charter for Mountain Regions has been prepared by the Council of Europe.[7] Both this and the Pan-European Biological and Landscape Diversity Strategy[8] highlight the needs for well-coordinated interdisciplinary research on environmental and societal change in mountain regions.

The European Commission's 1994–98 Environment and Climate Programme mentions mountain ecosystems as "of special importance", and has funded a number of projects focusing wholly or partly on mountain areas, including the Oxford conference in December 1997. The 5th Framework Programme for Research and Technological Development may also recognise the importance of mountain regions. Many European countries support significant levels of research on global change in mountain regions, either through national research funding agencies or through bilateral or multilateral aid.

3. European Conference on Environmental and Societal Change in Mountain Regions

The European Conference on Environmental and Societal Change in Mountain Regions took place in Oxford, UK, on 18–20 December 1997. Primary support was provided by the European Commission's European Network for Research on Global Change (ENRICH). Additional sponsors included international and European programmes, and British, French, and German organisations. This wide level of support permitted the participation of 128 individuals, from a very diverse range of backgrounds, from 30 countries on four continents. A total of 85 came from EU member states,[9] 18 from countries in transition,[10] 10 from developing countries,[11] and 15 from other countries.[12]

---

[6] Backmeroff, C., C. Chemini and P. La Spada (eds.) (1997) European Inter-governmental Consultation on Sustainable Mountains Development – Proceedings of the Final Trento Session. Guinta della Provincia Autonoma di Trento, Trento.
[7] Price, M.F. (1998) Chapter 13 in Action 1992–97- A Task Manager's Report. Food and Agriculture Organization of the United Nations, Rome.
[8] Council of Europe/United Nations Environment Programme/European Centre for Nature Conservation (1996) Pan-European Biological and Landscape Diversity Strategy. Council of Europe, Strasbourg.
[9] Austria, Belgium, France, Germany, Greece, Ireland, Italy, Netherlands, Portugal, Spain, UK.
[10] Czech Republic, Poland, Romania, Russia, Slovakia, Slovenia.
[11] China, Costa Rica, India, Nepal, Philippines, South Africa, Tanzania, Venezuela.
[12] Canada, Korea, Norway, Switzerland, USA.

Objectives of the Oxford conference:
1.   bring together individuals involved in natural and social science research relevant to global change in mountain regions, and those concerned with policy-making;
2.   review, summarise, and evaluate ongoing and proposed research activities with respect to global change in mountain regions;
3.   identify opportunities and mechanisms which address key issues for global change in mountain regions, emphasizing a) linkages between the natural and social sciences and policy-making and b) complementarity between programmes and initiatives;
4.   contribute to the ongoing scientific and policy agendas of the European Commission by making proposals for the future implementation of integrated, interdisciplinary research on global change in mountain regions;
5.   facilitate integrated networking of science and policy relating to global change in mountain regions, in Europe and at wider scales.

The programme was designed by an international and interdisciplinary Programme Advisory Committee, which met once before the conference, played a crucial role in its implementation, and subsequently drafted this final document. The content derives both from the presentations and from two series of workshops – regional and thematic – at the conference. The objectives of the workshops were to:

• review and summarise ongoing research;
• identify and discuss key research issues and opportunities and mechanisms to address them;
• propose priorities for integrated interdisciplinary research.

4. Needs and opportunities for integrated interdisciplinary research

The presentations and workshop discussions at the Oxford conference led to the identification of four sets of key issues for global change research in mountain regions:

• inventory and collection of baseline data on global change;
• research on processes of change in interacting environmental and societal systems;
• research on global change and mountain communities;
• implementation of research.

4.1. Inventory and collection of baseline data on global change

Inventories of existing data
A central issue to be resolved in the future implementation of global change research in mountain regions concerns the availability of data. For every potentially relevant variable, there is significant variation in the reliability, accuracy, and length of record at every scale. Only a few coordinated inventories have been undertaken. While a global review of mountain climate

data is a rare example[13], global data on key variables – for instance, the volumes of water flowing from mountain regions, or of the total number and distribution of mountain inhabitants – are not available. The relevant timeframe varies from the seasonal (e.g., land-use, land cover) to the decadal (e.g., human populations) to centuries (e.g., extreme events, such as major landslides and floods). Such inventories are essential in order to understand the dynamics of change, identify differences and commonalities between regions to permit agenda setting and prioritisation for research, and explore potential paths of change.

Increased attention needs to be given to information sharing, rather than the collection of new data. Similarly, in many regions, better topographic information is essential for the development of geographic information systems (GIS) and modelling of change. Regional institutions have a particular role to play in networking and in information compilation and dissemination.

One region for particular attention should be Central and Eastern Europe, where a wealth of data and experience regarding the environment and societies of the mountains of the region exists, some for long time periods. This provides an unusual opportunity for global change research; one priority should be the synthesis and comparison of historical information, which describes long trends whose future potential paths could be valuably projected according to various scenarios driven by various interactions of environmental and societal forces. This work should be focused by clear questions and hypotheses.

Monitoring key indicators
A basic necessity for global change research is the monitoring of key variables, each of which varies at distinct spatial and temporal scales. These variables describe characteristics of the biophysical, climatological, and socio-economic components of mountain systems; a key issue is how to link these sets of variables.

As recognised in the workplan for the IGBP inter-core project, emphasis should be placed on biophysical variables that are easily monitored and which relate to other environmental properties (e.g., snowline, glacier mass balance). The location of monitoring sites requires the resolution of a number of issues: including appropriate scale, the 'representativity' of river basins, and the needs, first, for regionalisation rather than measurement at a point and, second, to focus on key areas for sustained interdisciplinary research.

---

[13] Barry, R.G., (1992) Mountain climatology and past and potential future climatic changes in mountain regions: A review. Mountain Research and Development 12: 71–86. Price, M.F. and R.G. Barry (1997) Climate change B. Messerli and J.D. Ives (eds.) Mountains of the World: A Global Priority. Parthenon, New York/London: 409–445.

Key indicators for monitoring

- **Climatological/hydrological**: e.g., air temperature, precipitation, runoff
- **Cryosphere**: e.g., snowline, snowcover, extent of glaciers and permafrost
- **Landscape changes**: e.g., incidence of geomorphic phenomena (e.g., debris torrents, landslides, floods)
- **Biological**: e.g., species occurrence/abundance at high elevations, elevation of treeline, characteristics of high-altitude aquatic ecosystems
- **Economic**: e.g., monetary input/output, dependency ratio analysis, capital saving, equity
- **Social**: e.g., nutritional status, food security, health and hygiene, morbidity, empowerment (e.g., human resource development, capacity building, decision making), institutional arrangements for participatory decision making

---

Monitoring must be systematic, and undertaken both along altitudinal gradients and at the scale of entire river basins. Some monitoring may be best accomplished through remote sensing. In order to develop greater understanding of processes and potential paths of global change, all monitoring should be combined both with modelling, in order to sharpen thinking and estimate 'unobtainable' data, and with the collection of topographic and thematic (e.g., vegetation, terrain type) data. Recognising the paucity of existing climatic information, a clear priority is to maintain and develop systems of high-altitude weather stations and river gauges. In addition, global change indicator networks should be developed. All of these networks will require interdisciplinary approaches using a variety of methodologies at different scales.

Global change indicator networks

- glacier mass balance programmes[14]
- a global indicator network for the effects of climate change on high-elevation vegetation[15]
- comparative research on remote mountain lake ecosystems[16]
- information on landscape changes, especially the incidence of geomorphic phenomena, building on the experience of the International Decade for Natural Disaster Reduction

---

4.2. Research on processes of change in interacting environmental and societal systems

The driving forces of global change in mountain regions are interlinked and complex. Each of the four topics below considers aspects of global change at the interface between environmental and societal processes. It must be recognised

---

[14] Haeberli, W. and M. Hölzle (1993) Fluctuations of Glaciers, 1985–1990. World Glacier Monitoring Service, Zurich.

[15] Grabherr G., Gottfried, M., Gruber, A. and Pauli, H. (1995) Pattern and current changes in alpine plant diversity. Chapin, F.S. and C. Korner (eds.) Arctic and Alpine Biodiversity: Pattern, Causes and Consequences. Springer, Berlin-Heidelberg: 167–181.

[16] Patrick, S. (1997) MOLAR (Mountain Lake Research): project summary. Abstracts, European Conference on Environmental and Societal Change: 25.

that successful interdisciplinary research will require integration of these sets of issues.

Carbon and nitrogen cycles

The storage (sequestration) of carbon, as well as nitrogen cycling, in mountain ecosystems is a major global change issue in mountain areas. This is a politically-driven issue deriving from recognition of the potential implications of climate change. Better basic research is required on actual and potential volumes of carbon and nitrogen sequestered by both natural and managed ecosystems, along both land-use and altitudinal gradients. For informed and equitable policy decisions regarding agroforestry and plantations to provide carbon sinks, benefits to local communities (forest and wood products, foods, water retention, biodiversity) must be clearly defined.

Biodiversity and protected areas

Perhaps more so than in other environments, biodiversity in mountain areas encompasses both natural and cultivated species (agrodiversity). The preservation and enhancement of biodiversity in natural, semi-natural, agricultural, forest, and agroforestry ecosystems requires the close involvement of local people, who have specific perceptions of their interactions with these ecosystems and knowledge that can be used to ensure that mountain species and ecosystems may be managed for both local needs and wider benefits. The conservation of biodiversity through the designation and management of protected areas must provide local benefits (food, income from tourism, etc.) as well as the global benefits of species and ecosystem preservation. Thus, work on changes in the biophysical environment must be integrated with research on conflict resolution over resource rights and other issues, and on changing institutions to address rapid economic change, for instance in relation to tourism.

Economic forces are central in defining future paths of the status of mountain species and ecosystems. In many regions, the forces of globalisation are reflected by shifts in agriculture from a diversity of local species, each adapted to specific microclimates and local needs, to monocultures grown for export. Such shifts affect not only economic and health status of mountain people, but often the stability of mountain ecosystems. The need to direct economic forces towards a new balance between production and the provision of societal benefits (e.g., stable and attractive landscapes, the survival of endangered species) has begun to be recognised in the policies and activities of certain countries, the European Union, and international organisations. Further work on such topics, to build scenarios relating also to other diverse driving forces as climate change (especially changing frequencies of extreme events), migration, and the evolution of communications networks, is essential for sustainable development in the mountains.

Protected areas (national parks, forest reserves, etc.) should be a particular spatial focus for such work, linking not only to economic forces but to the need to ensure ecosystem stability in relation to changing land-uses and climate change. Worldwide, the need to involve local people in both ownership and management of the resources of these areas is being recognised. Local benefits from these resources need to be carefully redirected from direct exploitation (e.g., hunting, poaching) to sustainable activities, often including tourism, a major force of environmental and societal change in mountains around the world. The design and implementation of networks of protected areas, each designed appropriately with zones for different objectives, should therefore be a focus of global change research, building on existing initiatives such as the European Commission's Natura 2000 programme, the Pan-European Biological and Landscape Diversity Strategy, the Convention on Biological Diversity, and the Global Environment Facility.

Gradual and rapid changes in mountain landscapes
Mountain landscapes are highly dynamic, due to complex interactions of hydrological, ecological, pedological, climatic, socio-economic and, in some regions tectonic, forces. Two priorities for global change research should be:

- the dynamics and potential trajectories of changes in land-use/cover and water use, and the ways in which these are driven by processes of demographic and cultural change (e.g., in relation to migration, urbanisation), economic and political forces (e.g., land tenure, new economic activities; European and national policies), and constrained by climatic and ecological conditions;
- the magnitude and frequency of geomorphic events (e.g., landslides, runoff and flooding, erosion and associated changes in nutrient flows). Critical issues include the scale and response time of systems; sediment delivery ratios; and the definition of thresholds in changing environments. Since hydrological processes are central driving forces, research should also consider hydrological regimes, and the extent to which these are changing, and are likely to change. Since soil resources are fundamental to livelihoods, research should also consider soil erosion and nutrient dynamics, and the ways in which these are changing under human pressures. Finally, the implications of the various socio-economic driving forces (e.g., patterns of land tenure, access to resources, and migration) must be considered.

Applied research is needed on the impacts of land-uses and soil conservation techniques on erosion and hydrology. Associated research should consider the choice of appropriate technical measures for avoiding or mitigating the impacts of extreme events and for reducing the accumulated impacts of gradual change, and also coping strategies for addressing environmental uncertainty. Such work is particularly essential to underpin local institutions responsible for the management of natural resources and natural hazards. An important area for complementary research is to better understand the abilities of different levels of institutions to monitor and respond to global change; and to determine which scale/s is/are appropriate for particular types of change. This work should recognise fundamental changes of cultural attitudes within mountain peoples in relation both to economic development and urbanisation and to different physical manifestation of global change.

Climatic oscillations and extreme events
One major uncertainty in current understanding of the future evolution of the Earth's climate relates to changes in the frequency of climatic oscillations, such as El Niño events. At present, there exists only a vague and generalised set of ideas about the local consequences of these phenomena in mountain regions. Although they can be detected using global data sets, these phenomena must be seen in the context of the whole range of notable climatic events (drought, frost, hail, rain, floods, etc.) occurring in living memory, recognising that these events and their threats to life and harvests can have many origins.

The consequences of El Niño events, often extreme, vary between mountain areas, even within one country. They present considerable opportunities to develop models for understanding societies' interactions with the forces of global change. Using the current event as a starting point, a comparative study should be undertaken in mountain regions around the world to document consequences of, and responses to, El Niño phenomena. It should focus on:

- perceptions of climatic hazards over the past 50 years, to contextualise the role of El Niño among the diverse environmental hazards affecting mountain people;
- review of government and non-governmental policies and action plans and their perceived impacts at different levels;
- regional differences in perceived climatic hazards through analysis of climatic data and the sensitivity of policies and action plans to these differences.

4.3. Global change and mountain communities

Traditionally, mountain communities are connected to ecosystem functioning. Therefore, a close relationship exists between ecological and social processes in determining ecosystem and landscape integrity. To some extent, this has determined cultural characteristics and their relationships to their immediate environment and environmental uncertainties. The local-global interactions due to globalisation processes serve to integrate local economic forces and cultures. Mountain communities, generally located in economic peripheries, can create development activities which sustain the mountain environment, as well as their own social and cultural characteristics.

Mountain-lowland interactions
The mountains are important sources of many resources that are important for lowland populations. However, the concept of mountain-lowland interaction has not been strongly developed by scientists, who tend to focus on the processes of either mountains or lowlands. The value of the concept ranges from the policy- and decision-making level to socio-economic and cultural aspects. Policies and decisions concerning mountain regions are largely taken at national or even

international levels, with implications for a wide range of societal processes through influences on land-use, transport infrastructure, dams, and tourism developments, as well as agricultural practices, both through subsidy policies and the introduction of new technologies.

Mountain-lowland interactions

Impacts of and/or contributions **from the lowlands on the mountains**
- air pollution
- visitors, and the revenue and cultural impacts they bring
- subsidies/incentives (e.g., relating to agriculture, forestry)
- disturbance (e.g., landscape fragmentation, physical damage to soils, forest plantations)
- alien/introduced species, which often outcompete mountain species

Impacts of and/or contributions **from the mountains to the lowlands**
- water: freshwater, energy, and sediment transfer
- places for tourism, recreation and enjoyment
- timber and forest products
- landslide risks and other catastrophes
- emigration of people
- minerals
- agricultural products
- sacred sites

Mountain-lowland socio-economic interactions are significant, and are linked particularly through demographic processes. In mountain regions with low agricultural production, there is substantial migration, both seasonal and permanent, in order to compensate for a declining land/man ratio. In the opposite direction are large seasonal flows of pilgrims, recreationists, and tourists, some of whom become permanent immigrants. Both temporary visitors and immigrants have significant economic, social, and cultural impacts. An increasingly urbanised global society is having impacts not only directly on mountain cultures, but also on mountain ecosystems. Maintaining the diversity of mountain ecosystems requires understanding of the cultural diversity of mountain communities. There is a clear need for research on all of these issues of mountain-lowland interaction, with an emphasis on promoting sustainable production and consumption systems. Such research should be conducted at different spatial scales, in order to approach sustainable mountain development in a more comprehensive way.

Sustainable agriculture and forestry management and development
Traditional mountain agriculture and forest management has come under a whole variety of pressures, such as those related to intensification, and the globalisation of agricultural technology. A focus is needed on identifying and developing methods of sustainable landscape management which build upon traditional agricultural systems and innovative approaches to forest management involving the participation of local people. This is critical both for the

conservation of biodiversity and to ensure the stability and resilience of both natural and human-managed ecosystems. In turn, this demands designing appropriate agricultural and forest development strategies for the mountains and benefit-sharing mechanisms between mountain communities and society at large.

Economic and technological changes and mountain communities
Major technological and economic forces have deeply modified the economic bases of mountain communities. Many, especially in industrialised countries, have turned to economies based on a single activity, becoming more sensitive to economic cycles and various uncertainties. Three essential components of global change research in the mountains are to understand the evolution of these economic systems, to understand societal perceptions and their impacts on the environment, and to evaluate the capacities of mountain communities, in order to develop sustainable production systems in an uncertain world.

Since the economic bases and livelihoods of mountain communities have been transformed, it is important to understand how the identity of mountain communities is threatened, discussed and negotiated in relation to global economic and political changes (globalisation), and external social demands on mountain communities. In this respect, analysis of the complex phenomena of migration, particularly interactions between change in residential location and working place in mountain and nearby urban areas, is essential. An associated area for research is on the recognition and quantification of externalities relating to the management of mountain resources and the development of infrastructure. Such economic research is necessary both as a basis for the compensation of mountain people for ensuring the long-term provision of goods and services to wider communities, and to internalise the negative impacts of 'development' and engineering projects.

Appropriate institutional arrangements
In many mountain societies, institutional arrangements for natural resource management have traditionally been integrated. Under the impact of external pressures and associated processes of globalisation, a whole variety of sectoral institutions often have worked at cross purposes, leading to ecological and social disruption. Research is needed to provide a clear understanding of traditional institutional arrangements, the processes which contribute to changing them, and the alliances and interactions between mountain communities and interest groups at different levels. This knowledge is essential for developing appropriate institutional arrangements for the future. These local- and regional-level institutions should then be placed in the context of the developing global-level institutional mechanisms under the Convention on Biological Diversity, the Framework Convention on Climate Change, the World Trade Organisation, and other intergovernmental mechanisms.

4.4. Implementation of research

Interdisciplinary research is essential for global change studies because of the multiplicity of driving forces, both environmental and societal, of all aspects of global change in mountain regions. The aims should be to ensure dialogue, mutual understanding, sharing of information, and synergy, in order to have clear common research objectives and integrated implementation.

Methodologies

Once the availability of baseline data is known, a primary concern relates to methodologies for research on global change: decisions as to the scales at which new data should be collected, and how to nest data collection at different scales. For natural scientists, these range from plots, through transects along altitudinal gradients, to small headwater basins and large river basins. For social scientists, scales at which data are collected vary from households, to villages, census districts, and countries. A major issue for global change research is how to integrate the collection and analysis of data by different disciplines. These methodological issues, and the associated questions of how to handle data derived from investigation at different spatial and temporal scales, are central to the future definition of research. They will form a central part of the developing IGBP inter-core project.

Research partnerships and networks

A critical set of issues concerns optimal strategies for defining and implementing interdisciplinary research in mountain regions. Three needs may be identified:

- identification of experiences and barriers from past and ongoing studies and in existing disciplines and institutions, as has been done, for instance for the mountain projects within UNESCO's Man and the Biosphere (MAB) programme in Europe;[17]
- involvement of local people wherever possible in defining and implementing research, recognising the importance of local (traditional) knowledge and questions of equity. Partnerships between local people and scientists are essential to ensure the complementarity of local knowledge and scientific investigation regarding the causes and implications of environmental change. In research funded and implemented by scientists from industrialised countries in developing countries, local scientists should be partners; the research should utilise their local knowledge and expertise and be used to provide training;
- clear communication at all stages of research. At the definition stage, it can be used to determine the needs and wants of local people and other stakeholders. Near the end of research activities, it is necessary to transform, translate, and explain scientific results (information) into knowledge which is useful for local people and policy-makers at all scales.

---

[17] Price, M.F. (1995) Mountain Research in Europe: An Overview of MAB Research from the Pyrenees to Siberia. UNESCO, Paris, and Parthenon, New York/London.

Unless ecological, economic, cultural and political spheres interact, we will not reach a satisfactory means of resolving priority issues. For instance, unless environmental research is sufficiently targeted towards the main "brokers" in mountain/alpine environments, the prospects for the results of research being adhered to are poor.

5. Integrated networking of science and policy

As a complementary activity to the Oxford conference, the ENRICH programme is supporting the development of a database of ongoing research in mountain regions on themes relevant to global change. Entries to the database can be provided through a variety of means, including e-mail and the Internet. The database is accessible and searchable through the Internet, and outputs will also be available in printed form. Every effort is being made to ensure that it complements, rather than duplicates, other databases of scientists active in mountain research. On the Internet site, 'hot links' permit direct access to these and other relevant databases. Continued support for the functioning and updating of the database will be needed; this might best be done through the European Mountain Forum (EMF), part of the global Mountain Forum, "a global network in support of mountain people, environments and sustainable development". The secretariat of the EMF will be established during 1998 at the headquarters of IUCN-The World Conservation Union, in Switzerland.

An additional area for ongoing support is the development and linkage of databases of information concerning mountain regions. While a number of regional databases specifically consider mountain systems, the organisations responsible for global databases with relevant information should be encouraged to consider the applicability of this information for research on global change in mountain regions. Similarly, research and development funding agencies (national, European Commission) should draw attention to data resulting from projects that they have funded, and encourage funded scientists to make these data available through the Internet and by other means. For a concerted approach to be made towards better understanding and action relating to global change in the world's mountains, there is a clear need to link these databases through appropriate organisational structures. These should be managed through an international organisation, such as UNESCO, FAO, or the IGBP-DIS (Data and Information System).

Examples of databases relevant to global change in mountain regions

Regional databases; research activities

- Alps: Alpenforschungsinstitut (Garmisch-Partenkirchen, Germany)
- Alps: Swiss Academy for Natural Sciences (ProClim, Bern, Switzerland)
- Asia-Pacific region: (Asia-Pacific Mountain Network, at ICIMOD, Kathmandu, Nepal)

Regional databases; information

- Alps (Alpine Information and Observation System, Joint Research Centre, Ispra, Italy)
- Andes (INFOANDINA, implemented as part of CONDESAN)
- Himalaya-Hindu Kush (Mountain Environment and Natural Resources Information System, MENRIS, at ICIMOD, Kathmandu, Nepal)

Global databases; information

- Global Environment Monitoring System (UNEP)
- IGBP Data and Information System
- World Data Centre-A for Glaciology (Boulder, Colorado, USA)
- World Runoff Data Centre (Koblenz, Germany)
- World Precipitation Data Centre (Offenbach, Germany)

In addition to ensuring the availability of knowledge about mountain research and its results, structures for the cooperation of scientists in order to implement the various research activities outlined above are clearly needed. A number of regional frameworks for mountain research already exist, such as the Alpine Forum, ALPNET (supported by the European Science Foundation), ARTERI (supported by the European Commission), ICIMOD, and CONDESAN. The Mountain Forum also involves scientists from around the world. Another key set of structures is through the various global organisations and institutions that are increasing their attention to global change in mountain regions; a trend that should be encouraged. These include the International Geosphere-Biosphere Programme (IGBP), the International Human Dimensions Programme (IHDP), Diversitas, the International Geological Correlation Programme (IGCP), and Tropical Soil and Biological Fertility (TSBF). To support these actions, research and development funding agencies, and academies of science, should support elements of regional and global programmes on global change at a wide range of locations and altitudes, recognising the benefits of comparative research to the global community.

The greatest challenge is to develop more effective transfer of integrated knowledge and experience to policy-makers, local populations, and resource managers at global and regional scales. Appropriate means need to be found to take forward this agenda in future European Commission programmes within the context of the 5th Framework Programme for Research and Technological Development.

Bruno Messerli
Geographical Institute, University of Bern
Hallerstrasse 12, 3012 Bern
Switzerland

The global mountain problematique

During the conferences of Rio 1992 and Rio+5 (New York, 1997), the following question was very often heard from members of the political delegations: Do mountains really need their own chapter in 'Agenda 21'? Are these problems not much more national responsibilities? It may be true that mountains cover approximately one quarter of the global land surface (27% higher than 1000 m, 1% higher than 2000 m) (Ives et al., 1997b), but do they really have common problems and resources with a global significance for humanity? The answer is clear and it became even clearer during the special session of UN General Assembly in 1997: Three types of resources will play a fundamental role in the further implementation of 'Agenda 21' in the 21st century: water, biodiversity, and recreation for a rapidly growing urbanised world.

Water towers for the 21st century

All of the world's largest rivers originate in the mountains. Because they store immense amounts of fresh water as ice and snow, and in lakes and reservoirs, mountains play a global role in supplying water for agriculture, industry and urban centres in adjacent lowlands (Bandyopadhyay et al., 1997). In Europe, the Alps occupy only 11% of the total area of the Rhine's drainage basin, but they contribute between 50 and 60% of its flow in the summer months. More than 95% of the water in the basin of the Aral Sea comes from the Tien Shan and Pamir mountains.

Rivers also cross international boundaries and very often create political conflicts. Consequently, water management is highly complex, requiring technical, political and economic co-operation between upstream providers and downstream consumers. Co-operation will be ever more urgent in the 21st century, as 35% of the world's population is likely to experience water scarcity by the year 2025. Water management for the future begins in the mountains!

Mountains: Focal points of global biodiversity

Mountains are core areas of global biodiversity. Even after millennia of human modification of the Earth's ecosystem, mountain areas remain refuges for a great wealth of endemic species, specialised ecosystems, and remarkable landscapes (Jeník, 1997). Tropical and subtropical mountains, in particular, include major centres of global biodiversity (Barthlott et al., 1996). Many but not all of the mountain areas with the greatest biological and landscape diversity are protected areas. In fact, of the 785 million ha of protected areas in the world, one third are in the mountains, in 65 different countries (Thorsell, 1997).

The loss of the world's biological diversity continues. Urgent and decisive action is needed to conserve and maintain genes, species, and eco-systems. Recent advances in biotechnology show that genetic material in plants, animals, and micro-organisms has potential for agriculture, the health and welfare of human beings, and environmental protection. In this respect, mountain areas play a critical role in the sustainable use and long-term preservation of biodiversity on our planet.

Recreation for an urbanised world

More than half of the world's population lives in urban areas, and urbanisation will continue in the next century. This rapid increase in global urbanisation means that there will be great demand for leisure and recreation areas. Coastal regions and mountains have the most attractive potential in this regard. But coasts are characterised by warm temperatures and are linear; these features lead immediately to urban-like concentrations of people and buildings. Mountains offer a rich variety of vast, natural landscapes. If we take into account that tourism employs more than 200 million people, a tenth of the world's workforce, and that this number will even increase in the coming decades, it is clear that mountains and highlands will play a decisive role in tourism, the world's largest industry (Price *et al.*, 1997). Will it be possible to strike a balance between economic and ecological aspects and interests, for the benefit of both the local and the global population?

Mountain societies for the management of mountain resources

Sustainable mountain development and resource use is only possible by and for mountain communities. Mountain peoples' intimate knowledge of their environ-ment has produced diverse cultural behaviour with a common denominator: sensitivity to environmentally practicable action. Local knowledge and experi-ence are necessary to sustain ecosystems, livelihood, and production. However, they must be integrated cautiously with new techniques and research findings. Governments must be ready to make proportionately greater investments to train and educate people in mountain regions and to compensate the stewards of mountain resources.

Governments have great responsibility for these so-called peripheral regions and their vital resources. The decentralisation of power from capital cities to mountain regions, towns, and villages would be one of the best ways to empower mountain inhabitants, reduce costs, increase efficiency, and create the necessary interest in sustainable development and management of mountain resources (Ives *et al.*, 1997a). But when 26 of 48 wars and armed conflicts (in 1995) take place in or affect mountain areas (Libiszewski and Bächler, 1997), we shall not achieve the aim of peaceful management of mountain resources for the benefit of humankind.

Conclusion

We must realise that mountains are dynamic landscapes and susceptible to anthropogenically and naturally induced changes: probably much more sensitive than any other major environmental system. Therefore, mountains are by far the most suitable innovative research laboratories and indicators of climate and global change (Price and Barry, 1997).

The time from Rio 92 to Rio+5 has been a crucial and successful five years of rapidly growing mountain awareness. Nevertheless, the enormous obstacles facing sustainable mountain development in the 21st century will not be overcome unless the problems are recognised and addressed by local communities, national policy, and international co-operation.

REFERENCES

Bandyopadhyay, J., Rodda, J.C., Kattelmann, R., Kundzewicz, Z.W., Kraemer, D. (1997) Highland waters – a resource of global significance. In Messerli, B. and Ives, J.D. (eds), *Mountains of the World. A Global Priority*. Parthenon, Carnforth and New York, pp. 131–156.

Barthlott, W., Lauer, W. and Placke, A., 1996: Distribution of species diversity in vascular plants: towards a world map of phytodiversity. *Erdkunde* **50**: 317–327.

Ives, J.D., Messerli, B. and Rhoades, R. (1997a) Agenda for sustainable mountain development. In Messerli, B. and Ives, J.D. (eds), *Mountains of the World. A Global Priority*. Parthenon, Carnforth and New York, pp. 455–466.

Ives, J.D., Messerli, B. and Spiess, E. (1997b) Mountains of the world – a global priority. In Messerli, B. and Ives, J.D. (eds), *Mountains of the World. A Global Priority*. Parthenon, Carnforth and New York, pp. 1–15.

Jeník, J. (1997) The diversity of mountain life. In Messerli, B. and Ives, J.D. (eds), *Mountains of the World. A Global Priority*. Parthenon, Carnforth and New York, pp. 199–231.

Libiszewski, S. and Bächler, G. (1997) Conflicts in mountain areas – a predicament for sustainable development. In Messerli, B. and Ives, J.D. (eds), *Mountains of the World. A Global Priority*. Parthenon, Carnforth and New York, pp.103–130.

Price, M.F. and Barry, R.G. (1997) Climate change. In Messerli, B. and Ives, J.D. (eds), *Mountains of the World. A Global Priority*. Parthenon, Carnforth and New York, pp.409–446.

Price, M.F., Moss, L.A.G. and Williams, P.W. (1997) Tourism and amenity migration. In Messerli, B. and Ives, J.D. (eds), *Mountains of the World. A Global Priority*. Parthenon, Carnforth and New York, pp.249–280.

Thorsell, J. (1997) Protection of nature in mountain regions. In Messerli, B. and Ives, J.D. (eds), *Mountains of the World. A Global Priority*. Parthenon, Carnforth and New York, pp.237–248.

Tony Gross
Convention on Biological Diversity, 393 Rue St. Jaques, Office 300
Montreal, Quebec, H2Y 1N9
Canada

Biodiversity and sustainable development in mountain regions

The Secretariat of the Convention on Biological Diversity was delighted to be invited to take part in this Conference and to be given the opportunity to address the opening session on the linkages between biodiversity and sustainable mountain development, and the role of the Convention.

The negotiation of the Convention on Biological Diversity took place during the preparations for the Earth Summit, and the Convention was opened for signature in Rio de Janeiro in June 1992. The Convention came into force 18 months later, in December 1993, a remarkably short time for such international treaties. Four years later, it has 171 Parties, confirming that it is seen by the global community as an important and strategic instrument in the post-UNCED process of implementing transitions to sustainable development.

The objectives of the Convention are the conservation of biological diversity, the sustainable use of its components, and the fair and equitable sharing of the benefits arising out of the utilisation of genetic resources.

The Convention uses the term 'biological diversity' to mean the variability among living organisms from all sources including all terrestrial and marine and other aquatic ecosystem and the ecological complexes of which they are part. This includes diversity within species, between species and diversity of ecosystems. It recognizes the intrinsic values of biological diversity and the ecological, genetic, social, economic, scientific, educational, cultural, recreational and aesthetic values of biodiversity and its components. It also recognizes the importance of biological diversity for evolution and for maintaining life sustaining systems of the biosphere, and that biological diversity is being significantly reduced by certain human activities.

Parties to the Convention are required to identify and monitor biodiversity, and to identify processes and activities with adverse impacts. They are required to develop and implement biodiversity strategies and action plans and to integrate biodiversity considerations into sectoral and cross-sectoral planning. In contrast to other species- or sectoral-based environmental treaties, the Convention has adopted an ecosystem approach. These obligations on Parties – to integrate the conservation and sustainable use of biodiversity into national decision-making, and to adopt an ecosystem approach – together with the interlocking nature of the three objectives of the Convention mean that the

Convention is extraordinarily broad in its scope. These obligations imply creating the preconditions for sustainable development.

How therefore does the Convention relate to the sustainable development of mountain regions?

Mountain regions are among the world's most vulnerable ecosystems. Because of their three-dimensional character, mountains have great hetero-geneity of plant and animal communities within relatively short distances, due to altitudinal variation, compass orientation, variation in soils, dissection by valleys and gorges, and the dynamic nature of surface changes. This makes for high levels of genetic, species and ecosystem biological diversity. Not only is biodiversity extremely high, but so is endemism, as 'ecological islands' often develop on individual mountains. Mountain ecosystems are of extreme importance as biodiversity 'hotspots', as water reservoirs, and as outstanding cultural sites. Maintenance of biodiversity and cultural diversity are intimately linked. When biodiversity is eroded, indigenous culture and traditional knowledge, with the livelihoods they sustain, tend to be similarly eroded.

The Convention is representative of a new generation of multilateral environmental treaties which constitute frameworks for national action and contain provisions for new and additional financial resources to enable developing country Parties to meet the incremental costs of fulfilling their obligations under the treaty. In this regard, it requires that special consideration be given to the situation of developing countries with environmentally vulnerable areas, including mountain areas.

The Global Environment Facility (GEF), operating the financial mechanism of the Convention, is developing criteria for an operational programme to support developing country Parties with the conservation and sustainable use of mountain ecosystems. The two secretariats, of the GEF and of the Convention, on the basis of guidance given by the Conference of the Parties, are discussing priority areas for support. These might include support to: new protected areas or enlargement of existing protected areas; creation of conserv-ation corridors, along and up-and-down ranges; transborder protected area cooperation; protection of water catchments and of montane cloud forests.

However, the priorities to be identified will be premised on the need to guarantee 'ownership' of the activity in question by local stakeholders. Bottom-up development, based on local participation and management, is nowhere more essential than in mountain regions, which typically present high levels of significant cultural landscapes and where restoration and repair are often sorely needed.

When it meets for the fourth time in May 1998, the Conference of the Parties will take stock. It will have completed its first programme of work (1995–97). Decisions have been made on ways to implement the provisions of the Convention with respect to marine and coastal ecosystems and agrobiodiversity. Programmes of work on forest and inland water ecosystems will be approved. Important decisions will be taken on how to implement the provisions of the Convention on traditional knowledge. Work on monitoring, assessment and the development of indicators will continue. Parties will begin to consider key areas such as benefit-sharing, impact assessment and minimising adverse effects, public education and awareness. With nearly all Parties actively developing and implementing national biodiversity strategies and action plans, Parties will assess their progress in integrating biodiversity into other sectoral and cross-sectoral plans, programmes and policies.

In the new programme of work, to be established in May, there will undoubtedly be the recognition that the Conference of the Parties needs to address mountain ecosystems. Many countries have indicated this as a priority. It is likely to be taken earlier rather than later in the new work programme.

The Conference of the Parties undertakes the further elaboration of the provisions of the treaty and provides guidance on their implementation. This guidance applies to Parties in their national implementation, but equally to harmonising the policies and programmes of those other agencies and bodies – the United Nations and its specialized agencies, development banks and bilateral donors, scientific and research bodies, non-governmental organisations and community organisations – without whose involvement implementation of the Convention cannot take place.

The Conference of the Parties depends upon receiving sound scientific advice. It has, for this purpose, a Subsidiary Body on Scientific, Technical and Technological Advice, whose mandate is to provide timely advice relating to the implementation of the Convention. This is a multidisciplinary body of government scientific experts, and is developing into the authoritative body for the provision of scientific advice on biodiversity.

There is thus an opportunity and a challenge for the mountain scientific community. At some point soon, Governments will begin preparing for these scientific discussions which will provide the basis for future policy decisions of the Conference of the Parties. In the context of an international treaty with a broad scope and the ability to guide the policies and programmes of a wide range of other actors, such decision-making can have potentially powerful implications out there, on the ground, where biological and cultural diversity will be retained or lost.

Bernard Debarbieux
Institut de Géographie Alpine
Université Joseph-Fourier, Grenoble
France

Is "Mountain" a relevant object and/or a good idea?

Let us start by pointing out a well-known but odd paradox that crops up in our research activities: all researchers specialised in mountainous regions are convinced that scientific research on the Earth's mountains is a worthwhile activity; all of them seek to maximise their objectivity and use standard scientific methodology to investigate "mountains". All believe that the objects we are seeking to comprehend belong to a single category that justifies sharing and discussing the results of our work, whether it be in Oxford, Kathmandu, Chamonix or wherever. And yet, we find to our surprise that we have the greatest difficulty in defining what a "mountain" is and what the world's "mountains" have in common (see Messerli, in this volume, for a very recent and clever testimony).

These observations are not a prelude to a new definition of the concept of mountain. On the contrary, I would like to take advantage of this paradox to ask two major questions and to present an historical and epistemological reflection about the status of the word "mountain" in our scientific research.

Question 1: What exactly is the status of the word "mountain" in current scientific practice?

A great deal of research in the history of science and in cultural history (see Broc, 1969 and Schama, 1995 as very valuable analyses) has been devoted to the ways in which our forbears sought to understand mountains. All of them show that, in the overwhelming majority of cases, naturalists and geographers had a clear, preconceived idea as to what a mountain is; their attempts at definition systematically tried to identify objective criteria (borrowed from geology, climatology, biogeography, zoology, etc.) which would support their intuitive idea of what a mountain should be. And indeed, for many years they believed themselves to be successful in their endeavours.

As time went on, however, and our knowledge broadened, people were forced to give up these attractive but erroneous beliefs; it was realised that a mountain could not be identified by an objective trait or by a specific function. And so the task of defining what mountains were became more difficult. Generations of scientists have wrestled with this problem of definition, but have never succeeded in solving it satisfactorily (see for complex and courageous endeavours Veyret, 1962; Jeanneret, 1992; Bandyopadhyay, 1992 and, for a general perspective, Debarbieux, 1989).

It is not my intention to try and assess the various efforts, including those which have been made in the 20th century – even though those using new concepts such as ecosystems and connectivity do seem to have proved more useful. All I want to do here is to call attention to the fact that, for three centuries, the various attempts at defining "mountains" have been very similar: they all try to identify *a posteriori* objective criteria for defining "mountains" whose existence is already postulated on the basis of a deeply felt *a priori* intuitive concept. However, since we have not succeeded in producing an objective definition, we cannot see in the word "mountain" a category that has been deduced using scientific logic: what we have in fact is a pre-category which has arisen from our intuition. This pre-category has more to do, therefore, with the world of ideas than the world of objects.

Notwithstanding the foregoing, this has prevented neither our distant forebears nor our contemporaries from making substantial contributions to our understanding of the natural and social processes that take place in the regions which we refer to as mountainous. There can be no doubt therefore that the terms "mountain" and "mountainous" have proved extremely useful for expressing the everyday idea which we have of the world around us; they help us to observe that world objectively, but they do not enable us to define a set of natural objects. Consequently, I am of the opinion that we can very well sum up scientists' attitudes by paraphrasing Yves Bonnefoy, the French literary theorist, when he speaks of poetry, by stating: *"I can't say what a mountain is, but I can point one out!"*

This brings me to my second question:

Question 2: Is this scientific but yet intuitive conception of what "mountain" is so different from the conception of mountains by those who are not scientists?

Here again, a brief historical aside will help me to answer this question.

The idea that mountains constitute a special category of natural objects goes back long before the Natural History of the 18th century. This idea is a trait of our modern concept of what Nature is in the Western World (see for example the work done by Gibson, 1989, about Flemish painting of the XVIth century).

Moreover, some very impressive comparative studies in anthropology and religious history (see Samivel, 1962 and Eliade, 1969) have shown that the "mountain specificity" idea, though formulated in very various ways, is one found among peoples all over the world. Consequently, local perceptions and knowledge affect the uses (forestry, stock raising, social activities, etc.) that the people make of these areas, and affect their natural characters. Scientists are not the only ones to subordinate their mountain-related activities to this initial intuition of the "specificity of mountains".

This observation suggests that the perceptions that indigenous populations have of the mountainous areas they inhabit deserve the attention of scientists interested in "change in mountain regions". It suggests that, even when devoted to the analysis of objective facts and events, scientists should try to see things through the eyes of local people; to build into their research projects for understanding the mountains of the Earth the understanding that those who inhabit and model them have of their own environment.

This reasoning has implications that may seem simple and logical enough as long as one is working on remote, exotic mountains. It only takes a rigorous methodology to be able to scientifically analyse local peoples' perceptions of mountains, since these are different in so many ways from those of their scientific counterparts. But the reasoning I have presented also holds for "western" mountains. Here, however, the complexity of the analysis is greater. Whether we are talking about the Highlands of Scotland, the Alps or the Carpathians, the men and women who live in these areas, and who protect and develop them, are no longer guided by pre-scientific representations of their environment. In areas such as these, scientific representations of mountains (height, climatic measurement, ecosystems, geologic differentiation, etc.) are widespread in our society and are systematically reinforced by a range of institutions (Alpine clubs, environmental lobbies, etc.) and national policies. In sum, the evolution of "western" mountain areas in recent years has been profoundly affected by the representations of scientific origin which social actors carry, exchange or contest, depending on the particular situation in which they find themselves. It is thus most important that we scientists should investigate how naturalist and geographical concepts of mountains have been taken over by society at large.

Consequently, and without calling into question the value and usefulness of the research carried out on the mountains of the Earth hitherto, I believe we must admit that scientists have built a representation of the world's mountains which is not independent of those which went before, and one which is less and less independent of those residing in the minds of an ever-growing percentage of the men and women who are modifying these mountains today.

This conclusion clearly does not mean we have no more work ahead of us. On the contrary it opens up vast vistas of new research which I can only encourage us to explore.

REFERENCES

Bandyopadhyay, J. (1992) On the perceptions of mountain characteristics. *World Mountain Network Newsletter*, 7: 5–7.

Bozonnet, J.P. (1992) *Des Monts et des Mythes: l'imaginaire social de la montagne.* Grenoble, Presses Universitaires de Grenoble.

Broc, N. (1969) *Les montagnes vues par les naturalistes et les géographes de langue française au XVIIIe siècle.* CTHS, Paris.

Debarbieux B. (ed.) (1989) Quelle spécificité montagnarde? *Revue de Géographie Alpine,* **67**.

Eliade, M. (1968) *Traité d'Histoire des religions.* Paris, Payot.

Gibson, W.S. (1989) Mirror of the Earth: the "world landscape" in XVIth century Flemish Painting. Princeton Princeton University Press.

Jeanneret, F. (1989) L'alpinitié en Europe et en Nouvelle-Zélande. *Bulletin de la Société neuchâteloise de Géographie,* **32–33**, 63–83.

Samivel (1962) *Hommes, cimes et dieux.* Paris, Arthaud.

Schama, S. (1995) *Landscape and Memory.* New York, Alfred A. Knopf.

Veyret, P. and Veyret, G. (1962) Essai de définition de la montagne. *Revue de Géographie Alpine,* **50** pp. 5–35.

Martin F. Price
Mountain Regions Programme, Environmental Change Unit
University of Oxford, 11 Bevington Road, Oxford OX2 6NB
United Kingdom

Introduction: Programmes and Methods

Mountain research has a long tradition, and many important scientific advances have derived from research in mountain regions; both in individual regions and through comparative research. However, until the early decades of this century, individual scientists typically focused on their own specialisations. From the 1930s, mountain scientists – particularly in France, Germany, and the Soviet Union – increasingly recognised the inter-relationships of mountain ecosystems, and also of the people who inhabit them. These developments were first brought together in a coordinated way from 1973, when Project 6 of UNESCO's Man and the Biosphere (MAB) Programme on 'Impact of human activities on mountain and tundra ecosystems' began.

MAB-6 was the first international interdisciplinary research programme on mountain regions, with projects in the Andes, the Himalaya, many Alpine countries, and the Spanish Pyrenees (Price, 1995). This programme was central to the development of mountain science, which has continued to evolve through a series of international and interdisciplinary programmes (Ives and Messerli, 1990). Recently, there have recently been calls for a new 'interdisciplinary, intercontinental, and intersectoral discipline': montology (Rhoades, 1997; Ives et al., 1997).

The papers in this section present a number of programmes and methodological approaches for the study of mountain environments and those who live in and depend on them. Some of these programmes are based firmly in the natural sciences: for instance, the major research programme in Glacier National Park, USA described by Fagre and his co-authors. This region has almost no permanent human habitation, and therefore the emphasis is on 'natural' biogeophysical processes, albeit increasingly influenced by human actions, both intentional and inadvertent. Similarly, recognising that, at the global scale, high mountain ecosystems are some of those least influenced by human activities, Pauli and his co-authors propose a global network for monitoring the effects of climate change. This network would be part of the developing IGBP/IHDP initiative outlined by Becker in a paper which was updated as a result of a workshop which took place four months after the Oxford conference. This clearly recognises the importance of human-environmental interactions in mountain regions, and could provide a vital framework for future interdisciplinary mountain research. Within this, the approach presented by Winiger could be of significant importance.

The remaining papers describe various initiatives at different stages of completion. Kalvoda and Rosenfeld note the importance of natural hazards in mountain regions, as described in a book which will shortly be published. Aldrich presents a developing programme on tropical montane cloud forests, which are of great value for their biological diversity, as well as for providing important societal benefits. All of these papers are presented from various scientific perspectives; the oral testimony project described by Bennett takes the very different approach of allowing mountain people to speak for themselves. This is not just a counterpoint, but must be an important element of future mountain research, which should recognise not only the detailed knowledge of mountain people, but the need to ensure that research projects are designed and implemented with their involvement, and for their benefit: a perspective that is recognised in the developing IGBP/IHDP initiative, and is an increasing emphasis within European Union research programmes.

REFERENCES

Ives, J.D., Messerli, B. and Rhoades, R.E. (1997) Agenda for sustainable mountain development. In: Messerli, B. and Ives, J.D. (eds.) *Mountains of the World: A Global Priority*. Parthenon, London: 455–466.

Ives, J.D. and Messerli, B. (1990) Progress in theoretical and applied mountain research, 1973–1989, and major future needs. *Mountain Research and Development*, **10**: 101–127.

Price, M.F. (1995) *Mountain Research in Europe: An Overview of MAB Research from the Pyrenees to Siberia*. UNESCO, Paris, and Parthenon, Carnforth.

Rhoades, R.E. (1997) Pathways towards a Sustainable Mountain Agriculture for the 21st Century: The Hindu Kush-Himalayan Experience. International Centre for Integrated Mountain Development, Kathmandu.

Mark Aldrich
World Conservation Monitoring Centre
219 Huntingdon Road, Cambridge, CB3 0DL
United Kingdom

Developing a global database of tropical montane cloud forests

Introduction

Tropical montane cloud forests (TMCFs) are high on the list of the world's most threatened ecosystems, and it is widely believed that the majority of those that remain are small areas or remnant fragments of their original extent. Much of their value is related to their unique characteristics of biodiversity and endemism and the functions that they provide. However, despite their considerable value, these fragile habitats are under increasing threat from a wide range of sources, and immediate action is required to achieve the conservation of remaining TMCFs before any more of these rare and valuable habitats are lost for good.

Development of a global database

To date, relatively little is known about the location, extent, and condition of the remaining areas of TMCF at the global level. In particular, detail is required on the protection status, biological importance, socio-economic conditions, and current threats of each site. Although detailed information exists for some specific sites, it is widely scattered and often not generally available.

In response to these issues, in 1993 a TMCF Symposium was organised in Puerto Rico (Hamilton *et al.* 1996), with the aim of drawing together information on the global extent of TMCFs. In addition to producing regional maps showing general concentrations of TMCF, and a list of sites by region and country, the participants recommended that relevant information should be collected and analysed in a co-ordinated way in order to improve understanding of these ecosystems, and to ensure that reliable information is made available to a range of users for effective decision making. They proposed the development of a standardised methodology and format for collecting information, through a world-wide inventory of TMCFs, with textual, numerical, and spatial data stored on a centralised database at the World Conservation Monitoring Centre (WCMC).

Following the development of a draft framework for recording data, a pilot information gathering exercise was conducted with the help of more than 40 contacts in South and Central America. This was complemented with material gathered from a comprehensive review of information available through research at WCMC.

Results

Following an intensive period of research and compilation, a draft directory of TMCFs has been produced. A summary bulletin *"Tropical Montane Cloud Forests: An urgent priority for conservation"* has also been published in the WCMC Biodiversity Bulletin Series No. 2, (1997).

Table 1: Global Totals Total number of countries, total number of regions and sites per country with the number of sites protected

Region	Total N°. of countries	Total N° of CF regions	Total N° of CF sites	Total N° of sites with protection
Southeast Asia	14	63	228	114
Africa	21	57	97	38
Central America	6	25	120	36
South America	6	37	160	76
Global total	**47**	**182**	**605**	**264**

The global overview shows that a total of 605 TMCF sites in 47 countries have been identified to date. The highest concentration is found in Latin America, where 280 sites (46%) are found in only 12 countries: the majority in Venezuela, Mexico, Ecuador, and Colombia. In Southeast Asia, 228 sites have been identified in 14 countries, principally in Indonesia and Malaysia, and to a lesser extent in Sri Lanka, Philippines and Papua New Guinea. In Africa, 97 sites have been recorded in 21 countries.

In total, just under half the sites identified have an element of protection by being within protected areas classified as meeting Management Category I-VI criteria of IUCN-The World Conservation Union. Others may be under less formal protection as private reserves or in collaborative forest management agreements, although this information is not available at this stage. In Southeast Asia, 50% of sites identified are within IUCN protected areas, while Latin America has 44% (Central America 30%; South America 47%), and Africa has 39%. However, despite the fact that cloud forests in these areas are legally designated as "protected", in practice many are under some degree of threat and are continuing to become more fragmented and in some areas completely lost, at an alarming rate.

Conclusions

The production of this draft directory as the first output from the database represents an important initial stage in the development of a standardised methodology and format for collecting information on TMCFs on a site-by-site basis. It is hoped that WCMC can continue to provide a focal point for information exchange, through further data gathering and dissemination in a subsequent phase of work. As a next step, a planning and advisory workshop is planned, to include participants from each of the major regions, with the aim of reviewing the progress made to date and identifying the key priorities for further work.

REFERENCES

Hamilton, L., Juvik, J.O. and Scatena, F. (eds.) (1996). *Tropical Montane Cloud Forests.* Springer, Berlin.
WCMC (1997). Tropical Montane Cloud Forests: An urgent priority for conservation. Aldrich, M., Billington, C., Edwards, M. and Laidlaw, R. (Eds). *WCMC Biodiversity Bulletin* N°. **2**. World Conservation Monitoring Centre, Cambridge, UK.

Alfred Becker
Potsdam Institute for Climate Impact Research (PIK)
P O Box 60 12 03, 14412 Potsdam
Germany

IGBP and IHDP activities in mountain regions

Mountain regions occupy about one-fifth of the Earth's surface. They are home to approximately one-tenth of the global population and provide goods and services to about half of humanity. Accordingly, they received particular attention in "Agenda 21", endorsed at the UN Conference on Environment and Development (UNCED) in 1992. Chapter 13 of this document focuses on mountain regions, and states:

> "*Mountain environments are essential to the survival of the global ecosystem. Many of them are experiencing degradation in terms of accelerated soil erosion, landslides, and rapid loss of habitat and genetic diversity. Hence, proper management of mountain resources and socio-economic development of the people deserves immediate action.*"

Moreover, mountain regions often provide unique opportunities (sometimes the best on Earth) to detect and analyse global change processes and phenomena:

(1) Due to the often strong altitudinal gradients in mountain regions, meteorological, hydrological (including cryospheric), and ecological conditions (in particular vegetation, soil, and related conditions) change strongly over relatively short distances. Consequently, biodiversity tends to be high, and characteristic sequences of ecosystems and cryospheric systems are found along mountain slopes. The boundaries between these systems (e.g., ecotones, snowline, and glacier boundaries) experience shifts due to environmental change and thus can be used as indicators; some can even be observed at the global scale by remote sensing.

(2) Many mountain ranges, particularly their higher parts, are not affected by direct human activities. These areas include many national parks and other protected, "near-natural" environments. They may serve as locations where the environmental impacts of climate change alone, including changes in atmospheric chemistry, can be studied directly.

(3) Mountain regions are distributed all over the globe, from the Equator almost to the poles and from oceanic to highly continental climates. This global distribution allows us to perform comparative regional studies and to analyse the regional differentiation of environmental change processes as characterised above.

Accordingly, mountain regions are of particular significance for global change research, which must be intensified, collaborative, and coordinated.

A joint IGBP/IHDP initiative

Recognising the significance of mountain regions for global change research, two core projects of the International Geosphere-Biosphere Programme (IGBP) (Biospheric Aspects of the Hydrological Cycle, BAHC, and Global Change and Terrestrial Ecosystems, GCTE), together with the South Asian Committee of the Global Change System for Analysis, Research and Training (START), organised a workshop in Kathmandu, Nepal in 1996. This resulted in IGBP Report 43: "Predicting Global Change Impacts on Mountain Hydrology and Ecology" (Becker and Bugmann, 1997). Global change research in mountain regions was further addressed at the IGBP Congress in Bad Münstereifel, Germany, in 1996; and at a workshop on "Dynamics of Land-use/Land Cover Change in the Hindukush-Himalayas" organised by the IGBP/International Human Dimensions Programme (IHDP) Land-use and Cover Change (LUCC) core project in Kathmandu, Nepal, in 1997.

The outputs from these meetings, together with those from the Oxford conference, served as the basis for a joint IGBP/IHDP (BAHC, GCTE, LUCC, Past Global Changes [PAGES]) workshop on "Global Change Research in Mountain Regions" in Pontresina, Switzerland in April 1998. The participants emphasized the need for interdisciplinary environmental change research in mountain regions, involving both natural and social scientists. Thus, in addition to the IGBP and IHDP core projects mentioned above, the Institutional Dimensions of Global Change (IDGC) and Global Environmental Change and Human Security (GECHS) core projects of IHDP, as well as START and its regional programmes, should join the group of collaborators.

Integrated approach

It is proposed that the IGBP/IHDP Mountain Initiative meeting will apply an "integrated approach" for observing (detecting, monitoring), modelling and investigating global change phenomena and processes in mountain regions, including their impacts on ecosystems and socio-economic systems. Both environmental aspects – in particular land-use / land cover changes and climate change – and socio-economic aspects – in particular social, economic, and political driving forces and changes – as well as their complex interactions and interdependencies will be taken into account in their mountain-specific forms. The ultimate objectives of the approach are:

- to develop a strategy for detecting signals of global environmental change in mountain environments;
- to define the consequences of global environmental change for mountain regions as well as lowland systems dependent on mountain resources; and
- to develop sustainable land, water, and resource management strategies for mountain regions at local to regional scales.

It is understood that, at least in the coming decades, socio-economic changes are likely to be at least as important as environmental changes in mountain regions. The environmental changes may significantly threaten sustainable development in these regions, and both environmental and socio-economic changes may reduce the ability of these regions to provide critical goods and services to society in the mountains as well as downstream and elsewhere; for instance, in terms of water and energy supply, biodiversity, attraction to tourists, and measures to avoid or mitigate damaging effects of disastrous events (floods, debris flows).

Research activities

Many ecological and hydrological studies in mountain regions have, to date, concentrated on single sites or single small catchments, respectively. Little is still known about the behaviour of many ecological and hydrological processes along altitudinal gradients or about the downslope progression of upslope impacts. Therefore, one set of priorities is to study hydrological and ecological processes and their interactions along altitudinal gradients, with a particular emphasis on the influence of land surface characteristics, especially land-use and its changes. A second set of priorities derives from the increasing awareness of the special susceptibility of mountain areas to global change, discussed above, which makes them particularly suitable for studying the impacts of climate change on the hydrosphere and biosphere. A scientifically sound, feasible methodology including observations, manipulative experimentation, and modelling is clearly required. A final set of priorities is based on the recognition that mountain regions are eminently important for human communities living in the mountains as well as downstream. In many mountain areas, land-use pressure as well as the risk of natural disasters is increasing. Hence the relationships between global change and sustainable development are of key importance.

Accordingly, the IGBP/IHDP mountain research initiative will be structured around the following overarching themes:

1. Long-term observation systems to detect and analyse signals of global change, as the mountain-specific component of the Global Observing Systems;
2. Integrated modelling framework for analysis, vulnerability assessment and predictive studies, including the development of scaling and regionalisation methods for mountain regions;
3. Environmental-change-related mountain specific process studies, in particular along altitudinal gradients and in associated headwater basins;
4. Development of strategies to ensure sustainable development in mountain regions, and to avoid or mitigate damaging effects of disastrous events.

An implementation plan is under development. This will be distributed as a draft to the international science community for consideration, comments and suggestions.

REFERENCES

Becker, A. and Bugmann, H. (eds.) (1997) *Predicting Global Change Impacts on Mountain Hydrology and Ecology: Integrated Catchment Hydrology/Altitudinal Gradient Studies.* IGBP Report 43, IGBP Secretariat, Stockholm.

Olivia Bennett
Oral Testimony Programme, Panos Institute
9 White Lion Street, London, N1 9PD
United Kingdom

Mountain Voices: an oral testimony project by the Panos Institute

This is a project about the people who live in mountains, in all their complexity and, for the development planner, their awkward individuality. Over the last three years, the Panos Institute (Panos) has been working with local environmental, cultural and development organisations in selected mountain regions to collect and record the testimony of local people, and to communicate their own experiences, understanding and concerns about their social and physical environment.

Objectives

The objectives of the project are to explore the changing environment, culture and history of highland regions through the direct testimony of those who live there:

- through the dissemination of the resulting information – at local, national and regional levels – to generate greater understanding and debate about the changing environment in mountains and highland regions, and the global importance of these changes;
- to amplify the voices of more marginalised groups and individuals, whose views and experience are often key to the development of their communities, but who are too rarely consulted about its planning or implementation;
- to provide participating organisations and communities with a body of life stories and information, and the skills and experience to build upon such work in the future;
- to put them in touch with communities with similar concerns in different contexts, and so open up their opportunities to share information, experience and contacts, and support their ability to respond to change, development and outside influence.

Methodology

Interviews are carried out in local languages with and by a variety of ethnic and occupational groups. Collections have been gathered in the Himalaya (India and Nepal); the Andes (Peru); Mount Elgon (Kenya); the highlands of Ethiopia and Lesotho; southwest and northeast China; and Appalachia (USA). By the end of 1997, some 200 testimonies had been recorded, transcribed, translated, summarised and indexed.

Funding permitting, Panos will produce a book (in 1998) based on the whole collection, as well as many local-language publications based on the individual collections. For example, at least four booklets (in several languages) are planned from the India and Nepal testimonies. One of these will focus particularly on the theme of water, drawing on villagers' views, experience and memories of different systems of water management, and the impact of environmental and policy changes. The material should go some way to filling an acknowledged gap between government statistics and on-the-ground realities.

As an information-gathering exercise, this project does not pretend to be scientific, but it reflects a belief that data often need to be seen in a wider context which includes elements of history, culture, and social custom. Above all, it acknowledges that people's perceptions are as valuable as any more verifiable facts in gaining understanding of their societies and their priorities. Oral testimony can illuminate these factors, while pinpointing areas that may need more detailed research.

The project involves local people both as narrators and interviewers. Panos deliberately includes and trains some people for the interviewing teams whose access to education has been limited, and who would not normally have the chance to become involved in this kind of collaborative research. When this "capacity-building" process works, there is a greater sense of ownership by the communities involved, and more likelihood that the material will be used in ways that are valuable to them. Other local and national activities already planned include using the testimonies to make radio programmes, and as a basis for publications aimed at a range of audiences, including schools, adult literacy classes, and discussion groups and workshops involving donors, NGOs, government officials and the communities themselves.

Panos's local partners keep the original tapes and transcripts, while the archive of translations will be kept at Panos. Containing the views and experiences of 300 individuals, it represents a wealth of material – vivid, challenging, full of human detail and variety – to complement and illustrate other forms of research.

Some of the themes which are emerging from the testimonies are common to all communities; others are particular to certain areas. Topics include changes in the environment, agricultural practices, livelihoods, trading patterns and sources of income; relations with lowland and other communities; gender; the impact of resource extraction and development; energy; bio-diversity; land tenure; traditional customs and knowledge and the impact of modern education; the cultural and spiritual significance of mountains; and the cause and effect of changing social patterns, including migration.

Each collection is but a snapshot, and cannot claim to be representative of entire mountain groups. But such a range of individual voices does provide a remarkably comprehensive and vivid picture of highland societies, their changing physical and social environments, and their concerns for the future. At the heart of many of these concerns lies the question posed by an Indian villager: "Does the government want development of people in the highlands? Or development of people outside, based on what they can get out of the highlands?"

Daniel B. Fagre, Carl H. Key, Joseph D. White, Steven W. Running, F. Richard
Hauer, Robert E. Keane, and Kevin C. Ryan
USGS Biological Resources Division, West Glacier, MT 59936, USA and
University of Montana, Missoula, MT 59847, USA and
USDA Forest Service, Missoula, MT 59807, USA

Ecosystem dynamics of the Northern Rocky Mountains, USA

Mountain ecosystems worldwide are under stress because of increased demand for their resources and stronger external forcing factors, such as regional landscape fragmentation and global climate changes. Simultaneously, the global values of mountain systems are gaining wider recognition. As mountains become more geopolitically important, there is a greater need for informed resource management to maintain ecosystem integrity. Although all ecosystems are complex challenges to understand, mountain ecosystems are unique because their topographic diversity and steep environmental gradients lead to greater levels of biodiversity, ecological heterogeneity, and rates of endemism than in less spatially complex terrain.

We developed an integrated program of ecosystem modeling and extensive field studies to (1) understand and quantitatively estimate the major ecological processes of northern Rocky Mountain landscapes and (2) to use these capabilities to estimate responses to stressors, such as global climate change, or landscape-level disturbances, such as forest fires. The ability to predict ecosystem dynamics under different scenarios should support better management of mountain resources.

Our research was focused on Glacier National Park, a 4,078 km^2 wilderness in the northern Rocky Mountains of North America. Because this protected area is surrounded by National Forests, Wildernesses, and other parks, ecosystem changes can mostly be attributed to climate change, or other phenomena of global dimensions, rather than local causes. Some of the park's larger glaciers, for instance, have been reduced to 28% of their former extent over the past 140 years, presumably due to climate shifts (Key *et al.*, 1998). This documented glacial recession is an indicator of other ecosystem changes likely to be taking place.

To model ecosystem processes and changes in mountain environments, we further developed and tested the Regional Hydro-Ecological Simulation System (RHESSys) (Running *et al.*, 1989; Band *et al.*, 1991). RHESSys combines remote sensing, ecological modeling, and geographic information system (GIS) technologies to produce spatially-explicit estimates of various processes such as evapotranspiration and hydrologic outflows. A topographically-sensitive routing routine was incorporated to better distribute water dynamically through mountain watersheds. This improved estimates of stream discharge and provided insights into the sensitivity of the model to scaling issues (White and Running, 1994).

FIRE-BGC was created by combining a gap-phase succession model and a forest biogeochemistry model to estimate stand-level dynamics, accumulated carbon, and tree regeneration, growth and mortality (Keane *et al.*, 1996a). FIRE-BGC also explicitly modeled the successional response of the landscape to forest fires. Together, RHESSys and FIRE-BGC can reasonably predict the structure and composition of mountain forest communities and the daily rates of ecosystem processes at various spatial scales.

Model improvements and validation used six years of field data on mountain climatology, snow distribution, glacier activity, stream hydrology, aquatic biota, forest demographics, and soil respiration. For instance, over 3,500 snow measurements taken from a variety of slopes and aspects in two topographically diverse mountain watersheds (approximately 400 km^2 each) have correlated well with model estimates (Fagre *et al.*, 1997). Similarly, hydrographs from seven streams continuously monitored for five years compared well with those simulated by the models using snow water equivalence. In watersheds with remnant glaciers, however, higher observed values during late summer underscored the contributions of glacial meltwater to streamflow and the need to include this source in models of this mountain system. Modeled stream temperatures reflected observed values closely and can be tied to thermal regimes of aquatic organisms in mountain watersheds (Fagre *et al.*, 1997). Forest field measurements taken on 97 plots for three summers indicated that tree growth rates, vegetation type, soil depth and other carbon budget parameters were reasonably simulated (White *et al.* 1998).

Validated models have been used to assess potential impacts of climate change and other environmental stressors under different management scenarios. Future ecosystem characteristics, such as increased fuel loading and potential intensity of forest fires, have been simulated for 200 years using a climate-warming scenario. Increased net primary productivity and available nitrogen occur when current fire suppression policies are terminated (Keane *et al.*, 1996b). Alpine tree communities, however, become nitrogen-stressed, especially when interannual variability is increased in the future scenarios (White *et al.*, 1998). Upper and lower treelines shift upward and increased smoke emissions (and lower air quality) may become future management problems (Keane *et al.*, 1997). As more frequent fires occur, the forest composition shifts to disturbance-adapted species and to more deciduous species. Glacier National Park may lose its glaciers by 2030 (Hall, 1994) and basin water temperatures likely will increase sufficiently to alter stream macro-invertebrate communities. These and other results suggest that considerable potential exists for using simulation systems to project future changes to mountain landscapes. As both natural and human-caused changes accelerate, managers of mountain ecosystems should take advantage of these tools to guide important resource decisions.

REFERENCES

Band, L., Peterson, D., Running, S., Coughlan, J., Lammers, R., Dungan, J. and Nemani, R. (1991). Forest ecosystem processes at the watershed scale: basis for distributed simulation. *Ecological Modeling*, **56**: 171–196.

Fagre, D., Comanor, P., White, J., Hauer, R. and Running, S. (1997). Watershed responses to climate change at Glacier National Park. *Journal of the American Water Resources Association*, **33**: 755–765.

Hall, M. (1994). Predicting the impact of climate change on glacier and vegetation distribution in Glacier National Park to the Year 2100. M.S. Thesis, State University of New York, Syracuse, NY.

Keane, R., Hardy, C., Ryan, K. and Finney, M. (1997). Simulating effects of fire on gaseous emissions and atmospheric carbon fluxes from coniferous forest landscapes. *World Resource Review*, **9**: 177–205.

Keane, R., Morgan, P. and Running, S. (1996a). FIRE-BGC – A mechanistic ecological process model for simulating fire succession on coniferous forest landscapes of the Northern Rocky Mountains. INT-Research Paper–484, Intermountain Research Station, Ogden.

Keane, R., Ryan, K. and Running, S. (1996b). Simulating effects of fire on northern Rocky Mountain landscapes with the ecological process model FIRE-BGC. *Tree Physiology*, **16**: 319–331.

Key, C., Fagre, D. and Menicke, R. (In Press). Glacier recession in Glacier National Park, Montana. In Williams Jr., R. and Ferrigno, J. (eds.), *Satellite Image Atlas of Glaciers of the World*. United States Government Printing Office, Washington D.C.

Running, S., Nemani, R., Peterson, D., Band, L., Potts, D., Pierce, L. and Spanner, M. (1989). Mapping regional forest evapotranspiration and photosynthesis by coupling satellite data with ecosystem simulation. *Ecology*, **70**: 1090–1101.

White, J. and Running, S. (1994). Testing scale dependent assumptions in regional ecosystem simulations. *Journal of Vegetation Science*, **5**: 687–702.

White, J., Running, S., Thornton, P., Keane, R., Ryan, K., Fagre, D. and Key, C. (In Press). Assessing regional ecosystem simulations of carbon and water budgets for climate change research at Glacier National Park, USA. *Ecological Applications*.

Jan Kalvoda[1] and Charles Rosenfeld[2]
[1]Department of Physical Geography and Geoecology
Charles University, Prague, Czech Republic
[2]Department of Geosciences, Oregon State University
Corvallis, OR 97331, USA

Geomorphological hazards in high mountain areas

This paper presents brief information on the volume "Geomorphological Hazards in High Mountain Areas" (Kalvoda and Rosenfeld, In Press), prepared as a contribution of the International Geographical Union's Commission on Natural Hazards Studies (CONAHA). It is a report about CONAHA activities concerned with all serious hazards world-wide, with the aim of improving knowledge of the laws of hazards and risk estimation, thus leading to the possibility of their prediction and the reduction of the consequences on people. High mountain areas are among the regions where natural disasters with high risks are most frequent.

Several specialists were asked to contribute manuscripts according to the general theme of the volume, and a "correspondence symposium" resulted in the selection of manuscripts. These include selected regional examples of research into volcanic and post-volcanic activity, active fault zones, earthquakes, landslides, rockfalls, avalanches, creep movements, intensive weathering, erosion of soils and parent material, glacier thawing and surging, debris flows, mud torrents, floods and man-made landform changes. These phenomena, processes and events are described from various regions of the high mountains in the world, starting from the Mount Everest region in Nepal/Tibet and other High Himalayan and Karakoram mountain regions, from Tibet, China and Malaysia. Other contributions cover the Rocky Mountains in the USA, Mexico's volcanoes including Popocatepetl and Citlatepetl, the Peruvian and Bolivian Andes, and the European high mountains.

Contributions to the GeoJournal Library volume "Geomorphological Hazards in High Mountain Areas"

1. Prologue (J. Kalvoda);
2. Dedication to Professor Clifford Embleton (Ch. L. Rosenfeld);
3. Sedimentation and clast orientation of deposits produced by glacial-lake outburst floods in the Mount Everest region, Nepal (D. A. Cenderelli, E. E. Wohl);
4. Catastrophic flood flushing of sediment, western Himalaya, Pakistan (J. F. Shroder, Jr., M. P. Bishop, R. Scheppy);
5. Landslides and deserted places in the semi-arid environment of the Inner Himalaya (J. Baade, R. Mausbacher, G. A. Wagner, E. Heine, R. Kostka);

6. Glacier-induced hazards as a consequence of glacigenic mountain landscape, glacier and moraine dammed lake outbursts and Holocene debris production (M. Kuhle, S. Meiners, L. Iturrizaga);
7. Chaos theory of slides/mud flows in mountain areas. Example: Xiaojiang Basin, NE Yunnan, China (A. E. Scheidegger);
8. The salt weathering hazards in deserts (A. Goudie);
9. Impact of conversion of upland forest to tourism and agricultural land-uses in the Gunung Kinabalu Highland, Sabah, Malaysia (W. Sinun, I. Douglas);
10. Landslides in the Rocky Mountains of Canada (D.M. Cruden, X. Q. Hu);
11. Late Holocene sturzstroms in Glacier National Park, Montana, U.S.A. (D. R. Butler, G. P. Malanson, F. D. Wilkerson, G. L. Schmid);
12. Storm induced mass-wasting in the Oregon Coast Range, U.S.A. (C. L. Rosenfeld);
13. Natural hazards in relation to present stratovolcano deglaciation: Popocatepetl and Citlatepetl, Mexico (D. Palacios);
14. Andean landslide hazards (T.A. Blodgett, C. Blizard, B. L. Isacks);
15. Fluvial hazards in a steepland mountain environment, southern Bolivia (J. Warburton, M. Macklin, D. Preston);
16. Geomorphological response of neotectonic activity along Cordillera Blanca fault zone, Peru (V. Vilimek, M. Z. Luyo);
17. Geomorphological hazards and risks in the High Tatra Mountains (J. Kalvoda);
18. Geomorphologic hazards in a glaciated granitic massif: Sierra de Gredos, Spain (D. Palacios, J. de Marcos).

Geomorphological hazards studies in high mountain regions consider part of a larger group of natural hazards and are typical examples of research in which a combination of field research and theoretical interpretation is required. Disruption of the long-term stability of natural systems can pose a risk for all types of human activities. At present, many international research projects focus on determining the extent to which geodynamic processes affecting the Earth's surface constrain land-use and engineering.

The protection of people from natural risks in high mountain areas requires an understanding of the nature of hazards phenomena. Therefore, research into basic geomorphological hazards in high mountains can be viewed not only as a set of case studies, but as a rare opportunity for the preparation of theoretical models and for understanding the general architecture of the origins of natural disasters.

REFERENCE

Kalvoda, J and Rosenfeld, C. (eds), (In Press). *Geomorphological Hazards in High-Mountain Areas.* GeoJournal Library, Kluwer Academic Publishers, Dordrecht.

Harald Pauli, Michael Gottfried, and Georg Grabherr
Department of Vegetation Ecology and Conservation Biology
Institute of Plant Physiology, University of Vienna, Althanstrasse 14, 1091 Wien
Austria

A global indicator network for climate change effects on the vegetation in high mountain ecosystems: proposals from an Austrian IGBP/GCTE research initiative

Introduction

Ecosystems at the low-temperature limits of plant life are generally considered to be particularly sensitive to climate change. The option to use biocoenoses at the altitudinal limits as ecological indicators of climate change effects has attained increasing scientific interest (e.g., Beniston, 1994; Guisan et al., 1995; Price and Barry, 1997). One essential contribution was evidence for an ongoing upward shift of mountain plants in the Eastern Alps, provided by research within the framework of the Austrian IGBP (International Geosphere-Biosphere Programme) (Grabherr et al., 1994, 1995; Gottfried et al., 1994; Pauli et al., 1996).

This paper contributes to the implementation of the IGBP Mountain Workplan (Becker and Bugmann, 1997) (see Becker, this volume), and is particularly related to task 2.3 of activity 2 of the Mountain Workplan: Establishment of a global monitoring programme.

The indicator network

Accordingly, this contribution is a first outline on how to develop and establish a global indicator network to monitor responses of high mountain flora, fauna, and vegetation to climate change effects at different scales of time and space. The outline is based on the results of research within the framework of the Austrian IGBP/GCTE (Global Change in Terrestrial Ecosystems core project) in high mountain ecosystems of the Alps (see Gottfried et al., this volume).

The following objectives are suggested in order to meet the requirements of a successful and feasible global indicator network. The objectives are structured along the spatial scale, using the Stubaier Alpen (Central Alps, Austria) as an example.

Latitudinal arrangement of monitoring regions

At least one target region in a representative high mountain system within each major zonobiome (ZB; see Walter, 1985) along the latitudinal gradient has to be selected: ZB1: tropical-humid/equatorial; ZB2: tropical-humido-arid; ZB3: subtropical-arid; ZB4: mediterranean/arido-humid; ZB6: temperate/nemoral; ZB7: arid-temperate/continental; ZB8: cold-temperate/boreal; ZB9: arctic/polar. As a future option, this set of monitoring regions should be extended by additional target regions in transition zones between the major biomes.

Altitudinal arrangement of monitoring sites

The monitoring sites of each target region should be positioned at the most sensitive ecotonal boundaries, from the cold-determined treeline upwards. In addition, the sites should be largely untouched by anthropogenic land-use. In simplified terms, the sensitive high mountain ecotones of the study region in the Alps (nemoral zone) are: treeline, dwarf shrub boundary, grassland boundary, and limits of vascular plants.

Standardised general sampling procedure

Agreement must be reached on standardising the sampling procedure (focus, method, design). This is a prerequisite for comparing ecological patterns of different mountain regions. Two different, synergistic approaches – the Single Mountain and the Multi Summit approach – are suggested:

- The Single Mountain approach considers one mountain slope system per target region. It covers all major ecotonal boundaries from the treeline up to the limits of vascular plants. The main focus lies on permanent plot transect-studies at each of the sensitive ecotones in order to detect community patterns, ecological species groups or ecofunctional types along small-scale gradients caused by topography and local climatic influences. In addition, landscape and vegetation patterns along the entire slope system, as well as species richness in the transect areas, are to be recorded in a large-scale focus.
- The Multi Summit approach considers summits of different altitude (from treeline up to the limits of vascular plants) in the surroundings of the Single Mountain. This second approach is included to detect species richness and species assemblages along the altitudinal, and therefore the fundamental climatic gradient. The summit terrain is most suitable for an altitudinal comparison, because it includes all exposures within a small area.

Monitoring activities

The suggested monitoring activities within the frame of the Single Mountain and the Multi Summit approach are shown in Table 1 and Table 2, respectively. Activities and research focuses are structured along a spatial scale and a time scale for the intervals of re-investigation.

Table 1: Activities within the Single Mountain approach

Spatial scale	Activity	Focus	Interval of re-investigation
entire slope system	vegetation mapping	landscape patterns and large-scale vegetation patterns	50 to 100 years
100 × 100 m quadrats at the sensitive ecotones	habitat mapping, species mapping	habitat richness, species richness	5 to 10 years
2 × 2 m up to 3 30 m permanent plot transects within the quadrats	vegetation sampling	community patterns	5 to 10 years
1 × 1 m permanent plot units within the transects	vegetation sampling	species/ecofunctional types assemblage and abundance	5 to 10 years
0.5 × 0.5 m quadrats near the transects	experiments (e.g., temperature enhancement exp., see Molau, 1996)	species assemblage and abundance, phenology of key species	period of 3 to 5 years (no re-investigation)
individual species	laboratory analysis	ecophysiological envelopes of key species	no re-investigation
selected micro-habitats in different altitudes	temperature measurements (mini-dataloggers)	temperature gradient (e.g. soil temp. at 5cm, length of snow-free period)	continuous measurement

Table 2: Activities within the Multi Summit approach

Spatial scale	Activity	Focus	Interval of re-investigation
summit areas in different altitudes	species mapping	species richness	10 to 50 years
1 up to 10 m² single permanent plots of typical vegetation at the summits	vegetation sampling	species/eco-functional types assemblage and abundance	10 to 50 years
1 exposed point at each summit	temperature measurements (mini-dataloggers)	altitudinal temperature gradient, link to regional climate	continuous measurement

The outline shown above does not claim to be a finalised monitoring plan. It should be taken as a working concept for an interdisciplinary discussion among scientists dealing with climate change in mountain regions. We hope that the suggested objectives will contribute to initiating a close international co-operation to implement the required climate change indicator network.

REFERENCES

Becker, A. and Bugmann, H. (eds.) (1997) Predicting Global Change Impacts on Mountain Hydrology and Ecology: Integrated Catchment Hydrology/altitudinal Gradient Studies. *IGBP Report 43*, IGBP Secretariat, Stockholm.

Beniston, M. (ed.) (1994) *Mountain Environments in Changing Climates*. Routledge, London.

Gottfried, M., Pauli, H and Grabherr, G. (1994) Die Alpen im "Treibhaus": Nachweise für das erwärmungsbedingte Höhersteigen der alpinen und nivalen Vegetation. *Jahrbuch des Vereins zum Schutz der Bergwelt*, München, **59**: 13–27.

Grabherr, G., Gottfried, M. and Pauli, H. (1994) Climate effects on mountain plants. *Nature*, **369**: 448.

Grabherr, G., Gottfried, M., Gruber, A. and Pauli, H. (1995) Patterns and current changes in alpine plant diversity. In Chapin III, F. S. and Körner, Ch. (eds.), Arctic and Alpine Biodiversity: Patterns, Causes and Ecosystem Consequences. *Ecological Studies* Vol. 113, Springer Verlag, Berlin, Heidelberg, pp. 167–181.

Guisan, A. Holten, J.I., Spichiger, R. and Tessier, L. (eds.) (1995) *Potential Ecological Impacts of Climate Change in the Alps and Fennoscandian mountains*. Conservatoire et Jardin Botaniques de Genève.

Molau, U. (1996) ITEX climate stations. In Molau, U. and Mølgaard, P. (eds.), *ITEX Manual*, 2nd edition, Danish Polar Center, Copenhagen, pp. 6–10.

Pauli, H. Gottfried, M. and Grabherr, G. (1996) Effects of climate change on mountain ecosystems – upward shifting of alpine plants. *World Resource Review*, **8**, No. 3, Woodridge, Illinois, USA, pp. 382–390.

Price, M.F. and Barry, R.G. (1997) Climate change. In Messerli, B. and Ives, J. D. (eds.), *Mountains of the World: A Global Priority*. Parthenon, New York and London. pp. 409–445.

Walter, H. (1985) *Vegetation of the Earth and Ecological Systems of the Geo-biosphere*. Springer Verlag, Berlin, 3rd edition.

Matthias Winiger
University of Bonn, Institute of Geography
Meckenheimer Allee 166, 53115 Bonn
Germany

Mountain systems in transition: a methodological approach for the investigation of recent landscape dynamics

Changes of natural and human systems in mountain areas are complex phenomena of different temporal and spatial scales. They have been described, documented and analyzed in several case studies (e.g., the UNESCO Man and the Biosphere programme), sometimes including vertical gradients of processes, respectively highland-lowland interactions. But few examples consider monitoring approaches to land-use and surface-cover types. In consequence, there is no general database for comparative investigation on the dynamics of mountain landscapes within programmes such as "global change" or "human dimensions of global change". Analyses of this type generally apply geographic information system (GIS) techniques and therefore have to rely on strictly comparative data. Several basic reasons lead to constraints and deficits:

- inadequate or inhomogeneous geometric data base for most mountain areas;
- vast areas of poorly defined vegetation units, which in most cases are only extensively used for grazing;
- extent, type and status of forests are only randomly known;
- annual and long-term characteristics of the seasonal or perennial snow and ice coverage are available for only a few mountain ranges, although remote sensing techniques provide generally good results;
- morphodynamic patterns and processes (hazards) are monitored only very selectively;
- cultivated and settled areas as well as communication networks are eventually documented best in comparison to the previously mentioned aspects of the natural setting;
- in most cases periods of observation are short.

A methodological approach is presented and discussed, based on the combination of different types of data for spatial analysis of land-cover and land-use types. It is focused on its potential as a monitoring tool of visually recognisable land surface characteristics which are key factors for further topical analysis. It includes:

- ground-based analysis of vegetation, surface coverage and land-use types;
- density and age structure of forests;

- analysis of terrestrial photogrammetric (or photographic) data;
- satellite data (Normalized Difference Vegetation index [NDVI], classified multispectral information, seasonal snow cover);
- comparative evaluation of photographic data and thematic mapping covering a period of up to 100 years;
- digital elevation models (DEM) and GIS.

Results demonstrate that the dynamics in landscape changes are more complex and differentiated than generally assumed. Changes in forest cover, grazing and land-use intensity, morphodynamic activities etc. can be extremely diverse, and in most cases not at all uniform even within watersheds, an important fact for the understanding of controlling factors of the processes. Case studies include examples from the Karakorum, East Africa and the Alps.

Engelbert Ruoss[1] and Des B.A. Thompson[2]
[1] Natur-Museum, Luzern, Kasernenplatz 6, CH–6003, Luzern, Switzerland
[2] Scottish Natural Heritage, 2 Anderson Place, Edinburgh EH6 5NP, UK

Introduction: the Alps and the mountains of Northern Europe

The Alps and the mountains of Northern Europe present formidable challenges for resolving conflicts between environmental, developmental and societal changes. Europe's mountain regions are topographically variable, differing in latitude, altitude, climate, vegetation, and size. Historically, the economic conditions of all these regions were closely tied to the natural environment. Through the impact of mobility, migration and globalisation, mountain communities are changing rapidly. These regions are highly affected by external factors such as tourism, demand for agricultural products, industry and policy, all leading to changes in tradition and attitude. Drastic changes in the agricultural sector, such as farm abandonment and changes in land-use, lead to landscape changes and declines in biodiversity, resulting in increased natural hazards, forest fires, eutrophication problems, water pollution and soil degradation (Wiesinger). High mountain areas are often occupied by protected areas (national parks, nature reserves, etc.). The different legal regimes for protected areas are a mirror of the regionally characterized development (Brooks). In such regions, protection programmes and economic use have to co-operate intensively.

The Alps and the other mountain regions of Northern Europe differ in distance and obstacles. Compared to Northern Europe, high ecological/species diversity as well as variable land-use correlate to the small-scale mosaic of land in the Alps. The Alps are one of the most intensively studied mountain areas in the world. Research on the high risks of avalanches, flooding, landslides or runoff is of major interest in these mountains (see Cligniez and Manche, Collins, Weinmeister). They are characterised by a relatively dense population and a dense traffic system, strong regional organisation with high cultural diversity, high energy production (hydro-electric power plants), intensive forest management, as well as naturally high environmental stress. Most of the northern European mountain areas are less populated, have less traffic, and are not as structured topographically as the Alps.

The problems of civilisation in mountain areas are similar to those in regions situated in economic peripheries. Sustainable development depends considerably on the interaction between mountain and urban areas and the balance between tradition, modernisation, and migration (Warren, Thompson). Ironically, the general viewpoint of the mountains was traditionally based on the flatlander's perspective. In order to obtain a proper perspective of mountain areas, it is necessary to focus on the main issues as perceived by the mountain inhabitants and indigenous knowledge of the environment. To solve problems and to avoid conflicts, a well-coordinated, interdisciplinary research network is necessary. This should be the basis for further discussion on the main issues for the future.

The workshop involved discussion on research issues, required inter-disciplinary approaches, and network requirements based on the four main topics proposed in the presentation by Perrin-Sanchis (Table):

- diversity and dynamics of alpine/mountain zones
- territorial restructuring process
- stabilisation and control of environment dynamics
- social, cultural and political appropriation of dynamics

The participants in the workshop concluded that there seems to be an important need to set targets for restoring nature, embracing the full range of physical and biological features. The approach adopted by the Alpine Forum (Perrin-Sanchis) provides an important and valuable framework for implementing a systematic approach to analysing the issues and identifying ways of resolving them. It is quite clear that unless ecological, economic, cultural, societal, and political spheres interact, we will not reach a satisfactory means of resolving priority issues. Equally, unless environmental research is sufficiently targeted towards the main "brokers" in mountain/alpine environments, the prospects for success will be poor. Therefore it is important to separate the research issue from the interdisciplinary approach.

As ever, a number of surprising issues emerged from the contributions made by the workshop participants. First, there is still inadequate research information on nitrogen deposition and carbon balance in mountain environments. Second, we need to be cautious in recommending some technical measures which may become unsuitable after a certain period. For instance, some of the restorative techniques applied to steep gradient environments may, with time, prove to be only moderately effective. There was a resounding call for the creation of a mountain restoration network which would pull together researchers, practitioners, and those with an interest in economic and political matters.

Table 1: Analysis of issues relating to natural and human-influenced aspects of change in alpine/mountain environments

	1. Diversity and dynamics of alpine/ mountain zones	2. Territorial restructuring processes	3. Stabilisation and "control" of environmental dynamics	4. Social, cultural and political appropriation of dynamics
RESEARCH ISSUES	• Snow/glaciers/permafrost • Eco-hydrological systems • Water resources at high altitude • Bio-physical features palaeo-environmental reading) • Targets for restoring nature • Ecosystem resilience	• Share experiences ... share global achievements • Visions – local/national agendas • 'Downstream/Trans-frontier issues • Blend 'natural' and 'cultural' landscape issues	• Lifespan of technical measures • Steep torrent issues • What are we trying to sustain? • N deposition; carbon balance • Risk assessment	• How do locals value their areas? • Regional-National partnerships? • The power brokers? • Why do people leave the mountains? • Policies too much resources • Liability – insurance • "Modernisation"
INTER-DISCIPLINARY APPROACH	• Climate – landuse change • Key indicators of change • Interactions between disciplines	• Agricultural 'market' change and variability • Technical ⟷ traditional knowledge	• Objectives? Who sets them? • What is driving "control" • Role of sustainable development	• "Bottom-up" development • Ownership of development/protection • Local/tourist acceptance of issues • Who gains/pays for subsidies?
NETWORKS	• Alpine Forum to involve other regions – Scandinavia, British Isles – in time	• Combine 'hard' and 'soft' knowledge – information in transfer • Engage social scientists! • Rural/Mountain-urban migration • Land ownership ⟷ partnership • "Rural deficit"	• Cross-discipline language • Mountain restoration network • Values attached to features	• Social-ecological-political groups. Who are the brokers? • Institutional conflicts • Forum approach to resolving issues – be proactive

Vincent Cligniez and Yannick Manche
Cemagref – Division ETNA,
2 Rue de la Papeterie, BP 76, 38402 St Martin d'Heres
France

Risk analysis for natural hazards in mountain regions: application to snow avalanches

Dealing with natural risks is essential in the process of decision making for the development of mountain regions. In order to define a risk level, we present a theoretical model of risk assessment based on the intersection of hazard zones and vulnerability. First the hazard level is defined and described according to its probability and intensity. Then the vulnerability parameter is determined by introducing the elements – such as roads, houses and other places where many people usually stay – that can be reached by the natural phenomenon. Finally the risk is computed in order to produce a useful map for land-use planning.

Hazard level mapping

The aim of the study is to define the natural hazard in terms of its spatial extent and two main parameters: intensity and probability of occurence. We define a model with three levels of hazard (Table 1).

Table 1: Hazard level as a function of its probability and its intensity

	High probability	Medium probability	Low probability
High intensity	High	High	Medium
Medium intensity	High	Medium	Medium
Low intensity	Medium	Medium	Low

The probability parameter is easy to define. We consider the return period of the hazard, which may be one year, 10 years, or 100 years. The strength parameter is more difficult to define, and depends on the nature of the hazard. Our definition is based on the level of destruction and death such a hazard can create.

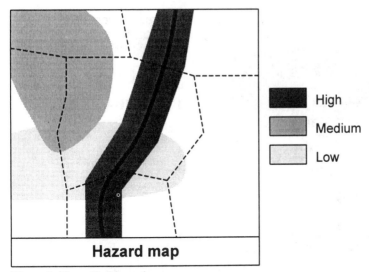

Figure 1: Theoretical hazard map

The hazard map is often considered as a risk map. One cannot conclude a given area is a risk zone because an intense probable hazard exists. A study of vulnerability is necessary to complement the hazard information.

Vulnerability studies

Vulnerability studies introduce the societal changes in mountain regions that lead to variations in risk, taking into account the persons and buildings that can be reached and damaged by the natural hazard (Manche, 1997). This parameter, vulnerability (Johnson *et al.*, 1994), is separated into two different aspects : direct and indirect vulnerability.

Direct vulnerability represents the zones, lines and points that can be reached by the phenomenon itself.

Indirect vulnerability represents the network features (roads, bridges, power lines, water pipes...) that, after being destroyed by the hazard, induce an indirect disaster, such as lack of access for emergency vehicles, electricity, heating energy or drinkable water.

Direct vulnerability is represented by shading, and indirect vulnerability is linked to the thickness of the network lines (Figure 2).

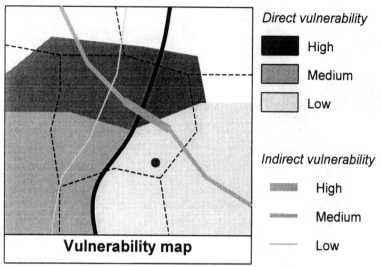

Figure 2: Theoretical vulnerability map

Risk assessment and management

With the above definitions of hazard and vulnerability, we are able to assess risk as a combination of these two parameters (Panizza , 1987).

Table 2: The risk definition as a function of hazard and vulnerability

	High vulnerability	Medium vulnerability	Low vulnerability
High hazard	High	High	Medium
Medium hazard	High	Medium	Medium
Low hazard	Medium	Medium	Low

This provides a spatial treatment of the information, to show locations, (zones, lines and points) that are in situations of both hazard and vulnerability (Figure 3).

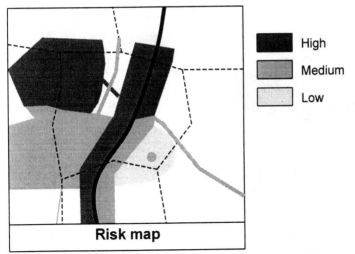

Figure 3: Theoretical risk map created with hazard and vulnerability maps

The white zones must not be considered as free of risk. In fact, people can be injured anywhere by something unpredictable. Those zones should therefore be interpreted as having an extremely low risk.

Application to snow avalanches

To create the risk map we have used a Geographical Information System (GIS) called ARSEN, dedicated to natural hazard modelling (Buisson *et al.,* 1995). This tool is based on the properties of spatial data structures for environmental studies (Worboys, 1992; Beattie *et al.,* 1996) (Figure 4).

There are other methods of risk assessment analysis for snow avalanches (Smith *et al.,* 1997), such as the avalanche-hazard index (Schaerer, 1989).

Conclusion

The consideration of vulnerability in risk studies is now possible with the simple model presented here. Because of the lack of information in many fields, such as hazard intensity and probability, the risk definition is restricted to three levels: high, medium and low. This precision is generally enough for decision making.

This paper has focused on early research. As natural risks are increasingly studied and represented for mountain development purposes, the extension of theoretical and practical mapping will improve methods in future decades.

Figure 4: Screen copy of risk application dedicated to snow avalanches

REFERENCES

Beattie, B. *et al.* (1996). Spatial Reasoning for Environmental Impact Assessment NCGIA Third International Conference on Integrating GIS and Environmental Modeling, Santa Fe, New Mexico, USA, January 1996.

Buisson, L. and Cligniez, V. (1995) Spatial Knowledge Base for Natural Hazards Protection : the ARSEN project. *Safety Science*, special issues TIEMES Volume **20**, pp. 27–37.

Johnson, G. O. and Reynolds, L.A. (1994). Using GIS for Hazards Vulnerability Analysis. Proceedings of The International Emergency Management and Engineering Conference (TIEMEC), pp. 3–8, Hollywood Beach, Florida, April 18–21 1994

Manche, Y. (1997). Propositions pour la prise en compte de la vulnérabilité dans la cartographie des risques naturels prévisibles. *Revue de Géographie Alpine*, N°2 Tome 85, pp. 49–62.

Panizza, M. (1987). Geomorphological hazard assessment and the analysis of geomorphological risk. *International Geomorphology* 1986, Part I, edited by V. Gardner, pp. 225–229.

Schaerer, P. (1989). The Avalanche-Hazard Index. *Annals of Glaciology*, n°13, pp. 241–247.

Smith, M. J. and McClung, D.M. (1997). Avalanche frequency and terrain characteristics at Rogers' Pass, British Columbia, Canada. *Journal of Glaciology*, Vol. **43**, N° 143, pp.165–171.

Worboys, M.F. (1992). A generic model for planar geographical objects. *International Journal of GIS*, Vol **6**, n°5, p 353–372.

David N. Collins
Alpine Glacier Project, School of Geography
University of Oxford, Mansfield Road, Oxford OX1 3TB
United Kingdom

Rainfall-induced high-magnitude runoff events in highly-glacierised Alpine basins

High-magnitude runoff events in rivers draining from glaciers and glacierised basins in Alpine areas result from both extreme hydrometeorological conditions and sudden outbursts of water stored in marginal ice-dammed lakes or in sub-/en-glacial pockets. Although outburst floods can be of considerably greater magnitude than those which are climatically determined, they are characteristic only of certain glaciers. The latter are, however, ubiquitous, and weather conditions leading to periods of sustained high ablation rates or to sustained precipitation events affect broad swaths of Alpine terrain.

Floors of Alpine valleys to which rivers descend down steep slopes from highly-glacierised high-altitude basins have been inundated on several occasions during the warm summers of the 1980s and 1990s, as a result of sustained rainfall during passage of storms over the European Alps. In 1987, floods resulting from heavy rain over the Swiss Alps between 14 and 19 July, and between 23 and 25 August, produced annual maximum flows with recurrence intervals in the range of 10–30 years for unregulated rivers draining partially-glacierised basins (Landeshydrologie und –geologie, 1988; Rey & Dayer, 1990; Zeller & Röthlisberger, 1988). Flooding was also catastrophic in Lombardia, Italy, in mid-July 1987 (Chardon, 1990). Floods with about 20-year recurrence intervals occurred in Alpine areas of Switzerland as a result of torrential rain between 22 and 24 September 1993 (Landeshydrologie und –geologie, 1994). On 5–6 November 1994, intense rainfall to high elevation gave rise to widespread inundation in Piemonte, north-west Italy, and elsewhere in the southern Alps (Pangallo, 1995). Storms in the Mont Blanc massif produced floods on the Arve at Chamonix, France, on 24–25 July 1996.

Analysis of the floods generated from Findelengletscher and Gornergletscher, Pennine Alps, Wallis, Switzerland in August 1987 (Collins 1995; in press) indicates the critical role of the transient snowline in the hydrology of Alpine glacierised basins, separating partial areas which are either (1) snow-free and return both rain and meltwater as runoff rapidly or (2) snow-covered, in which rain and melt percolate into snow to be delayed in transit or retained, according to thermal conditions in the pack. With respect to storm rainfall, the position of the transient snowline interacts with basin hypsometry to define the portion of the basin that will rapidly form runoff. Elevation of the 0°C isotherm also interacts with basin hypsometry to determine the area over which precipitation falls as rain. The quantity of water contributing to rain-induced augmentation of flow is related to the amount of rain falling during a storm and to the partial area of basin which is both snow-free and receiving liquid precipitation. Additionally, precipitation amount normally increasing with elevation will also influence runoff, the effect being greater the higher the transient snowline, the larger the snow-free catchment area, and the higher the freezing level in the atmosphere. The two adjacent glacierised basins exhibit differing responses to rainfall because of differences in both hypsometry and absolute elevation range.

Whilst the quantity of moisture delivered by a storm is the prime determinant of the runoff response, the timing of a storm in relation to the steady seasonal progression of the rise of the transient snowline and the more volatile episodic rising and falling of the freezing level during the ablation season is also critical in Alpine glacierised basins. The snowline is likely to reach maximum elevation and the snow-free area to become largest in the period from mid-August to mid-September. Delivery of a large quantity of precipitation in that window is therefore likely to be in the form of rain over an expanded non-retentive basin surface. Consequently, rainfall-induced floods are most likely to occur in late summer, following a period of elevated air temperatures. In warm summers after winters in which snow accumulation is below average, transient snowline and freezing level will rise sooner and higher, and extend the window of risk forward into July. In such warm summers, a given rainfall event might be expected to produce higher-magnitude runoff.

Hydrological conditions within glaciers at the time of storm impact also influence the magnitude of rain-induced floods from glacierised basins. Rainfall contributions added to pre-existing high melt-induced discharges, following periods of warm weather, will produce larger flood peaks. Rain over glaciers at times when the capacity of the internal hydrological system to discharge water supplied from the ice surface is low – early in the season before the drainage system expands, or in late summer when ice overburden pressure has closed conduits under recession of flow conditions – may lead to temporary storage of water and high subglacial water pressure. Runoff from rainfall might then trigger a sudden outburst, as appears to have been the case at Findelengletscher in September 1993 and at the Mer de Glace, in the Arve basin, in July 1996. Attempts to forecast floods from glacierised Alpine basins will require frequently updated information concerning elevation of the transient snowline; continuous air temperature measurement at high-altitude stations; continuous monitoring of water storage in glaciers, using water levels in boreholes; and close-interval measurement of rainfall intensity at high-elevation sites.

REFERENCES

Chardon, M., 1990: Les catastrophes naturelles de l'été 1987 en Lombardie crues, inondations, écroulement de Val Pola, *Revue de Géographie Alpine*, 78: 59–87.

Collins, D. N., 1995: Rainfall-induced high-magnitude runoff events in late summer in highly-glacierised Alpine basins, Proceedings of the British Hydrological Society Fifth National Hydrology Symposium, Edinburgh, 1995, 3.55–3.59.

Collins, D. N., in press: Rainfall-induced high-magnitude runoff events in highly-glacierised Alpine basins, International Association of Hydrological Sciences Publication, 248.

Grebner, D. and Richter, K. G., 1991: Gebietsniederschlag; Ereignisanalysen 1987 und Abhangigkeitscharakteristiken, Mitteilung der Landeshydrologie und –geologie, 14: 23–40.

Landeshydrologie und –geologie, 1988: Hochwasserereignisse im Jahre 1987 in der Schweiz, Mitteilung: 10.

Landeshydrologie und –geologie, 1994: La crue de 1993 en Valais et au Tessin, Mitteilung: 19a.

Pangallo, E., 1995: Alluvione del 5–6 novembre 1994 in Italia NW: analisi meteorologica, *Nimbus* 6–7: 13–21.

Rey, Y. and Dayer, G., 1990: Crues de l'été 1987 dans les bassins versant glaciaires des Alpes Pennines, *Revue de Géographie Alpine*, 78: 115–124.

Zeller, J. and Röthlisberger, G., 1988: Umwetterschäden in der Schweiz im Jahre 1987, *Wasser Energie Luft*, 80: 29–42.

Thomas Dirnböck and Georg Grabherr
Department of Vegetation Ecology and Conservation Biology
Institute of Plant Physiology, University of Vienna, Althanstrasse14, 1091 Wien
Austria

Vegetation mapping as a tool for estimating the hydrological balance of Alpine landscapes: the catchments of the drinking water resources of Vienna as an example

Introduction

The headwaters of high mountain systems are important water resources in many parts of the world (Messerli and Ives, 1997). As an example, the city of Vienna acquires about 95% of its drinking water from the karst mountains of the most north-eastern Alps. The catchment areas of these water sources are exposed to dynamic land-use change, and probably will be significantly affected by climate change (Grabherr *et al.*, 1995). In the framework of a long-term research initiative of the City of Vienna, a research project has been established to quantify how these changes might affect the amount and the quality of available water in the future.

As the basic approach, vegetation mapping was applied to get a quantitative figure (in terms of area occupied) of different ecological units with distinct hydrological characteristics and soil types. Vegetation mapping on the landscape scale was optimised by applying modern tools of data processing and data display: geographic information systems (GIS) and digital elevation models (DEM). Furthermore the relationships between floristically defined mapping units to soil features and hydrology will be shown.

The catchment areas of the drinking water supply of Vienna are typical mountainous karst systems in limestone and dolomite, with altitudes up to 2300 m. The plant cover changes drastically along the altitudinal gradient. Mixed forests with beech, fir and spruce, monodominant spruce forests, krummholz, various calcareous alpine grasslands, rocky slopes and screes are dominant types in different vegetation belts (Greimler and Dirnböck, 1996).

Water balance estimation: methodology

Vegetation has an essential effect on the spatial distribution of water balance parameters. Canopy characteristics, species abundance and temporal variation are important variables in soil-plant-atmosphere interactions (e.g. Kelliher *et al.*, 1993; Körner, 1994). In a rough way, vegetation maps are therefore areal representations of water balance. Each plant community has specific

43

hydrological properties, with interception and transpiration by the vegetation cover itself as well as the soil water properties. Based on the vegetation map, 36 ecotopes (functionally homogenous vegetation-soil-units) were defined.

To obtain estimates of evapotranspiration rates, a comprehensive literature survey of experimental research (e.g. Körner *et al.*, 1989) on similar vegetation types in the Alps was carried out. Soil water properties were measured in the field and assigned to the ecotopes.

The data sets (vegetation characteristics, species abundances and spatial distribution, mapped soil properties, DEM, estimates of evapotranspiration) were then connected to generate a preliminary water balance model. Again, the GIS represents the spatial database as well as the central methodological instrument.

An adjustment of the evaporation model was necessary in order to incorporate specific climate conditions of the study area. It was evident that the data collected from various experimental sites showed deviations due to distinct climate. Average precipitation, for example, of the most north-eastern Alps is comparatively low. One tool for adjustment was the Turc-Wendling Potential Evaporation Model (Wendling *et al.*, 1991). For this purpose, the percentage evapotranspiration values of the preliminary model were transformed into absolute rates, using precipitation distribution data. Comparing these to the potential evaporation rates, rough deviations and false estimations became visible and were iteratively adjusted. Further adjustment was through the catchment runoff contribution from springflow data.

Conclusions

1. Even in difficult alpine areas with pronounced relief patterns, vegetation mapping can be performed in a reasonable time when using GIS-based mapping tools.
2. Vegetation models relating relief patterns to vegetation types can be elaborated for expanding mapping to large areas and predicting future vegetation patterns determined by environmental and/or land-use change (Austin *et al.*, 1994; Fischer, 1994; Gottfried *et al.*, 1997).
3. At least for the Alps, many hydrological investigations have provided enough data for characterising vegetation units from a hydrological point of view (hydrological envelopes).
4. Vegetation maps are therefore valuable tools for establishing hydrological balances on a landscape scale. Due to the fact that vegetation-soil-units can be changed by human influences, such maps can play an important part in watershed management.

REFERENCES

Austin, M. P., Meyers, J. A. and Doherty, M. D. (1994). Predictive models for landscape patterns and processes, Sub-project 2, Modelling of landscape patterns and processes using biological data. Division of Wildlife and Ecology, Commonwealth Scientific and Industrial Research Organisation, Canberra.

Fischer, H.S. (1994). Simulation der räumlichen Verteilung von Pflanzengesellschaften auf der Basis von Standortskarten. Dargestellt am Beispiel des MaB-Testgebiets Davos. Veröff. Geobot. Inst. ETH, Stiftung Rübel, 122.

Gottfried, M., Pauli, H., And Grabherr, G. (1997). Prediction of vegetation patterns at the limits of plant life: a new view of the alpine-nival ecotone. (in review: *Arctic and Alpine Research*, Boulder, Colorado).

Grabherr, G., Gottfried, M., Gruber, A., Pauli, H. (1995). Patterns and current changes in Alpine plant diversity, *Ecological Studies*, **113**: 167–181.

Greimler, J. and Dirnböck, T. (1996). Die subalpine und alpine Vegetation des Schneebergs, Niederösterreich, Vegetationskarte im Maßstab 1:10000 und Beschreibung der Vegetation, *Linzer biologische Beiträge*, **28** (1): 437–482.

Kelliher, F. M., Leuning, R., Schulze, E.-D. (1993). Evaporation and canopy characteristics of coniferous forests and grasslands, *Oecologia*, **95**: 153–163.

Körner, C., Wieser, G., Cernusca, A. (1989). Der Wasserhaushalt waldfreier Gebiete in den österreichischen Alpen zwischen 600 und 2600 m Höhe, Veröffentlichungen des österreichischen *MaB-Programms*, **13**: 181–153.

Körner, C. (1994). Leaf diffusive conductances in the major vegetation types of the globe, *Ecological Studies*, **100**: 465–490.

Messerli, B. and Ives J. D. (eds.), (1997). *Mountains of the World – A Global Priority*, Parthenon Publishing, New York – London.

Wendling, U., Schellin, H.-G, Thomä, M. (1991). Bereitstellung von täglichen Informationen zum Wasserhaushalt des Bodens für die Zwecke der agrarmetereologischen Beratung, *Zeitschrift für Meteorologie*, **41**: 468–475.

Murray Ferguson[1] and Janet Adamson[2]
[1] Scottish Natural Heritage, 17 Rubislaw Terrace
Aberdeen, AB10 1XE Scotland, United Kingdom
[2] Cairngorms Partnership, 14 The Square
Grantown-on-Spey, Morayshire, PH26 3HG Scotland, United Kingdom

The Cairngorms: challenges of managing a Scottish mountain landscape

The Cairngorm mountains and the surrounding valleys or "straths" contain some of Scotland's most treasured wild landscapes and rare wildlife, as well as being home to some 20,000 people. Many more people come to visit the area to take part in a wide range of outdoor activities and, over several generations, the mountains have inspired many writers, poets and naturalists.

The core of the mountains has two high areas: the Cairngorm massif, containing the highest summit, Ben Macdui (1309 m), where the catchments of the Rivers Spey and Dee meet; and, to the south and east, the Mounth, a range of hills between the Cairngorm massif and the lowlands of Perthshire and Angus to the south. These mountains, which form the eastern extremity of the Scottish uplands, contain the largest area of ground above 1000 m in the UK. Around the high ground, particularly to the north and east, there are extensive areas of heather moorland and, in the glens, enclosed hill farmland.

The Cairngorm mountains are widely recognised as one of the most important areas in the UK for earth science and nature conservation. They are of international importance because of their arctic-alpine habitats and extensive montane blanket bogs, and the range and juxtaposition of landforms. The area contains many rare vascular and lower plants, some of which are only found in the Cairngorms. The lower slopes hold the largest remaining fragments of Caledonian pinewoods in the UK. The forests, slopes and plateaux support an important and distinctive assemblage of breeding birds, including the spectacular golden eagle along with ptarmigan, snow bunting and dotterel.

The mountains are vitally important for the surrounding local communities. Many jobs are generated by those who visit the area to enjoy an exceptionally wide range of recreational opportunities which include walking, skiing, rock and ice climbing, deer stalking, grouse shooting and nature study. The central parts of the mountains have a strong wild character which adds a quality to the recreational experience that is unique within the UK. The rivers and streams flowing from these hills are important for salmon or trout fishing and for a range of water sports. Furthermore the rivers and lochs are significant source of potable water, either "straight" or as a key ingredient in one of the region's most popular exports – whisky.

Management

Scotland does not have the legislative powers to designate National Parks but, in 1948, the Cairngorms were declared a National Park Direction Area which gave the area slightly stronger protection from proposed development. In 1954, the core area of the mountains was designated as a National Nature Reserve. Since that time, several other protective designations have been put in place. There has, however, been widespread and growing concern that the area, most of which is privately owned, requires better protection and more effective management to conserve its special qualities. Key issues of concern to environmentalists include the degradation of the woodland resource (principally because of ecologically unsound red deer populations); inappropriate and poorly controlled burning of moorland; the construction of bulldozed tracks; persecution of raptors; the impacts of tourism and recreation on sensitive habitats and species; the degradation of fish stocks; and airborne pollution arising from outside the immediate area.

In 1991 the Government established a Working Party to advise on the way forward. The findings were published in "Common Sense and Sustainability" (Cairngorms Working Party, 1992), which recommended that a Partnership of public, private and voluntary sector interests in the area should be established to produce a management strategy for the area.

Implementing these recommendations in 1994, the Secretary of State for Scotland appointed an independent chairman and provided funds through Scottish Natural Heritage to establish the Cairngorms Partnership. A Board representing the main interests in the area was convened, supported by an executive organisation with a small number of staff. The Partnership was charged with the preparation of a Management Strategy:

> *"to guide and encourage the use and management of land and other relevant activities in a manner compatible with both the environmental considerations and the need to promote the social and economic well-being of local communities"* (Secretary of State for Scotland, 1994).

Following the synthesis of existing information, three key documents were produced:

- The Cairngorms Assets: a statement of the principal features of interest, an analysis of threats, issues and concerns and identification of gaps in the information (Bayfield and Conroy, 1996);
- The Vision for the Future: a visionary statement of how the Partnership would like to see the area looking 25 years ahead (Cairngorms Partnership, 1996);
- The Policy Cascade: a synopsis of the existing range of policies and legal and other instruments and their implications of the management of the land (Tyldesley *et al.* 1996).

Following an extensive consultation process, "Managing the Cairngorms: The Cairngorms Partnership Management Strategy" was published in December 1997. The strategy is based on sustainable development, the precautionary principle and a voluntary approach, and attention is now being turned towards its implementation.

The Cairngorms Partnership will have a co-ordinating role in taking forward the principles outlined in the Management Strategy, while the partner organisations and individuals will be responsible for its implementation. Scottish Natural Heritage, for example, will continue to take the lead in the conservation and enhancement of the wildlife, habitats and landscapes of the Cairngorms and will encourage people to enjoy the area responsibly.

Challenges for the future

Many challenges remain for the management of the Cairngorms mountains. Amongst these the most significant are likely to be:

- how to win the support of the people who live and work in the area for the Cairngorms Partnership's Management Strategy;
- the rationalisation of "top down" processes (such as the commitments given by the Government to habitat and species protection under European legislation or the UK Biodiversity Action Plan), with the "bottom up" processes concerning the empowerment of local communities and the satisfaction of their aspirations;
- the debate that is currently underway concerning the Government's proposals, announced in September 1997, to consider designation of the area as one of Scotland's first National Parks;
- how to co-ordinate the efforts of the key partners (including five local authorities, many statutory organisations and voluntary bodies across a wide range of interests) to implement the Cairngorms Partnership's Management Strategy.

REFERENCES

Bayfield, N.G. and Conroy, J.W.H. (eds) (1996) The Cairngorms Assets, Cairngorms Partnership, Grantown-on-Spey.
Cairngorms Partnership (1996) The Vision for the Future Cairngorms Partnership, Grantown-on-Spey.
Cairngorms Partnership (1997) Managing the Cairngorms: The Cairngorms Partnership Management Strategy, Cairngorms Partnership, Grantown-on-Spey.
Cairngorms Working Party (1992) Common Sense and Sustainability: A Partnership for the Cairngorms, The Scottish Office, Edinburgh.
Secretary of State for Scotland (1994) Scottish Office Rural Framework: Cairngorms Partnership: A Statement of Intent by the Secretary of State for Scotland following the Advice of the Cairngorms Working Party, The Scottish Office, Edinburgh.
Tyldesley, D. et al. (1996) The Policy Cascade: The Cairngorms Partnership Legislative, Policy and Planning Framework, Cairngorms Partnership, Grantown-on-Spey.

Rob I. Ferguson and Owen C. Turpin
Department of Geography & Centre for Earth Observation Science
Sheffield University, Sheffield, S10 2TN
United Kingdom

Hydrological modelling in mountain environments

Runoff from glaciers and seasonal snowpacks is a vital water resource in many parts of the world, particularly for hydropower and arid-zone agriculture. An ability to forecast meltwater runoff helps water resource planning and has potential for assessing the implications of climate change scenarios. Some computer models have been developed specifically for snowmelt or glacier hydrology, and several general-purpose hydrological models contain snowmelt routines. Models differ in detail, but all involve four steps: (1) extrapolation of meteorological data to the snowpack, using temperature and precipitation lapse rates, (2) estimation of local snowmelt rates using some approximation of the energy balance, (3) integration of snowmelt over the snow-covered area (SCA), and (4) routing meltwater and rainwater to the basin outlet allowing for time-lags and evapotranspiration loss. The third step is critical in mountain environments because both accumulation and melting of snow are spatially variable, leading to a progressive decline in SCA over many months. Good predictions of runoff rely on accurate specification of SCA depletion, either as an input based on remote sensing or by simulating snowpack growth and depletion, using meteorological data.

The HYDALP project (Hydrology of alpine and high latitude basins), funded by the European Commission, is evaluating the operational use of remote sensing to assist runoff modelling in mountain environments. The project involves the Universities of Innsbruck, Bern, and Sheffield; the Swedish Meteorological and Hydrological Institute; and the Macaulay Land-use Research Institute in Scotland. Two very widely used meltwater models, HBV and SRM, are being used to simulate and forecast runoff in contrasting basins in the Austrian and Swiss Alps, the Scottish Highlands, and Arctic Sweden.

SRM (Martinec 1975; Martinec *et al.* 1994) is a relatively simple model specifically for snowmelt. It requires SCA time series (snow depletion curves) as input. The original approach (Rango and Martinec, 1982) was to establish a family of depletion curves from satellite images for previous melt seasons with contrasting winter snow depths, then to use images early in a new season to decide which curve to use. The preferred approach now (e.g. Nagler and Rott, 1997) is to input SCA in near real time, taking advantage of semi-automated image geocoding and classification. Either way, research is needed on the best way to interpolate SCA between satellite overpasses and the optimum frequency of input data in cost-benefit terms.

HBV (Bergström and Forsman, 1973; Bergström, 1992) is a more complex general-purpose model that currently makes no use of remote sensing since it simulates accumulation as well as melt. Satellite snow data at intervals during the melt season have obvious potential for testing the accuracy of the simulated spatial pattern of snow accumulation and melt. In the event of a discrepancy, parameters can be re-calibrated or snow depths updated.

Neither HBV nor SRM was designed specifically for basins containing glaciers, in which SCA depletion reveals fast-melting ice instead of bare ground. Models like HBV and SRM can be applied to glacial basins by specifying a higher melt rate for bare ice, but this requires knowledge of when and where ice is exposed. A glacier runoff model has been developed at Sheffield that simulates snowline rise using assumptions about the spatial distributions of initial snow water equivalent (SWE) and of melting (Turpin *et al.* 1997). The model has been calibrated on discharge data from the Findelengletscher in Switzerland and applied to the nearby Gornergletscher with re-calibration of glacier-specific parameters only. Verification tests on both glaciers show excellent runoff predictions ($R^2 > 0.85$). Simulated snowline elevations tend to be lower than estimates from Landsat data, but compensating over-prediction of icemelt generates good discharge predictions. Remote sensing thus gives insight into model behaviour, parameter errors, and possible improvements.

REFERENCES

Bergström, S. and Forsman, A. (1973) Development of a conceptual deterministic rainfall-runoff model. *Nordic Hydrology* 4: 147–170.

Bergström, S. (1992) The HBV model – its structure and applications. SMHI report RH4, Swedish Meteorological & Hydrological Institute, Norrköping.

Martinec, J. (1975) Snowmelt runoff model for river flow forecasts. *Nordic Hydrology* 6: 145–154.

Martinec, J., Rango, A. and Roberts, R. (1994) Snowmelt runoff model (SRM), user's manual, (ed.) Baumgartner, M.F., *Geographica Bernensia*, P29. Department of Geography, University of Bern.

Nagler, T. & Rott, H. (1997) The application of ERS–1 SAR for snowmelt runoff modelling. *Int. Assoc. Hydrol. Sci. Publ.* 242:119–126.

Rango, A. and Martinec, J. (1982) Snow accumulation derived from modified depletion curves. *International Association of Hydrolical Sciences Publication* 138: 83–90.

Turpin, O.C., Ferguson, R.I. and Clark, C.D. (1997) Remote sensing of snowline rise as an aid to testing and calibrating a glacier runoff model. *Phys. Chem. Earth*, 22: 279–283.

J.E. Gordon[1], V.M. Haynes[2], I.C. Grieve[2], V. Brazier[1] and D.B.A. Thompson[1]
[1] Scottish Natural Heritage, 2 Anderson Place
Edinburgh EH6 5NP, United Kingdom
[2] Department of Environmental Science, The University of Stirling
Stirling, FK9 4LA, United Kingdom

Change in mountain environments: geomorphological sensitivity to natural processes and human activity in the Cairngorm Mountains, Scotland

The Cairngorms

As oceanic, mid-latitude mountains with continental affinities, the Cairngorms represent an area of western Europe that is potentially highly sensitive in terms of geomorphological and ecological responses to environmental change and human activities. The Cairngorms comprise the largest continuous area of high ground above 1000 m in the British Isles, the nearest present-day analogue to an arctic landscape in Britain. The area is outstanding for nature conservation, and is recognised to be of national and international importance for a range of interests, including geomorphology, woodlands and montane (alpine) environments (Gordon, 1993; Thompson et al., 1994; Brazier et al. 1996). The climate reflects a unique combination of oceanic and continental influences, and the high montane (alpine) zone displays climatic, geomorphological and ecological similarities to parts of western Norway.

Sensitivity to change

The principal geomorphological features on the Cairngorm high plateaux include a variety of relict and active periglacial landforms which are strongly linked to the degree of exposure, the duration of snow-cover and the distribution of vegetation communities (Haynes et al., 1998). This study examines the sensitivity of these landforms to the main types of change which can be envisaged from human activities and climate change (Gordon et al. in press). Pressures from human activities on the environment of the high Cairngorms may arise from two types of source: 1) local land-use and human activities (e.g., recreation and grazing animals); and 2) global processes (climate change and atmospheric pollution). This distinction is important since the former tend to be more localised and may be more readily addressed by local management responses. The latter, however, may potentially have much more widespread effects, producing greater changes, but requiring management responses to be co-ordinated at national or even international levels. The properties of the regolith and soils are fundamental to an understanding of landscape sensitivity since they determine thresholds for change within the force/resistance relationships of the geomorphological system. In upland areas, processes are often episodic, being dependent on extreme events, so that information is also required on the magnitude and frequency of forces which are capable of overcoming soil/regolith resistances.

Environmental controls on processes were investigated in similar geomorphological contexts: exposed summits, cols, north-trending spurs, 'medium-lie' snow-patches and active patterned ground. The most sensitive sites are on steep slopes, e.g., leading down into cols or on the flanks of spurs, and in hydrologically active areas with deep regolith. Accessibility means that the slopes around Cairn Gorm and its surrounding cols and spurs are particularly vulnerable to human disturbance by trampling. The plateau regolith is deficient in the most erodible grain sizes, which reduces its sensitivity to many processes, especially compared to the more organic soils at lower altitudes. Processes are consequently intermittent and slow, and higher-magnitude events relatively important. Frost-susceptible regolith is widely present within reach of surface freezing.

Timeframe of changes

Truncated podzols were found to be widespread on the plateaux and cols, indicative of a period of greater stability in the past over all except the most severely exposed locations, followed later by erosion. These events are not dated, though the soil characteristics suggest soil development over timescales of an order of magnitude of 10^3 years and a period of disturbance lasting for centuries. The distribution of eroded areas is best explained by climatic exposure, so climatic deterioration seems likely to be a more important cause than anthropogenic influence, though the role of increased deer grazing cannot be entirely discounted. Recycling of eroded sand and gravel currently occurs across the ground surface in features such as 'vegetation islands', wind-stripes and possibly terracettes, which may be in dynamic equilibrium. In cols, there is frequently a clear boundary between vegetated areas, which have remained stable over the long term, and areas of truncated soils and patchy vegetation. These boundaries have clearly shifted in the past, and are potentially sensitive at present.

Monitoring needs

Long-term monitoring is required: 1) to investigate long-term rates of processes, which should help to distinguish the effects of natural variability and anthropogenic impacts; and 2) to provide early signals of significant changes. Pilot baseline surveys have been initiated, using a combination of ground survey and photography, with digital storage of data. This work is being developed as part of a formal, integrated monitoring programme for the Cairngorms, one of the first of its kind in the UK.

Conclusion

Although the present geomorphological systems and ecosystems have evolved through periods of both warmer and colder climate than at present during the Holocene, suggesting a long-term robustness, additional pressures now exist in the form of human activities. These may produce significant additional stresses that may ultimately increase the geomorphological sensitivity and reduce the biodiversity of the Cairngorms. The properties of the regolith and soils, combined with long-term climate-related changes, have exerted major influences on the patterns and processes of erosion and rendered some features much more vulnerable to present-day human influences.

REFERENCES

Brazier, V., Gordon, J.E., Hubbard, A. and Sugden, D.E. (1996) The geomorphological evolution of a dynamic landscape: the Cairngorm Mountains, Scotland. *Botanical Journal of Scotland,* **48**, 13–30.

Gordon, J.E. (1993) The Cairngorms. In Gordon, J.E. and Sutherland, D.G. (eds.), *Quaternary of Scotland.* Chapman & Hall, London. pp. 259–276.

Gordon, J.E., Thompson, D.B.A., Haynes, V.M., MacDonald, R. and Brazier, V. (In Press) Environmental sensitivity and conservation management in the Cairngorm Mountains, Scotland. *Ambio.*

Haynes, V.M., Grieve, I.C., Price-Thomas, P. and Salt, K. (1998) The geomorphological sensitivity of the Cairngorm high plateaux. *Scottish Natural Heritage Research, Survey and Monitoring Report*, No. 66.

Thompson, D.B.A., Horsfield , D., Gordon, J.E. and Brown, A. (1994) The environmental importance of the Cairngorms massif. In Watson, A. and Conroy, J. (eds.), *The Cairngorms. Planning Ahead.* Kincardine and Deeside District Council, Stonehaven. pp.15–23.

M. Gottfried, H. Pauli, K. Reiter and G. Grabherr
Department of Vegetation Ecology and Conservation Biology
Institute of Plant Physiology, University of Vienna, Althanstrasse 14, A–1091 Wien
Austria

The Austrian research initiative: global change effects at the low temperature limits of plant life: a progress report

Our research activities, which focus on changes of vegetation patterns at the low-temperature limits of plant life, are structured along three methodological approaches: empirical observations based on revisitations and monitoring, experiments simulating climate change, and scenario modelling.

Empirical observations

Revisitations of 30 subnival and nival summits of the Alps, at altitudes above 3000 m, indicated that the high alpine flora is currently moving upward (Grabherr et al., 1994, 1995; Gottfried et al., 1994; Pauli et al., 1996). These summits, first visited 50 to 100 years ago by botanists such as Braun-Blanquet (see, e.g., Braun, 1913; Reisigl and Pitschmann, 1958), can be regarded as the oldest permanent plots for vegetation monitoring in the Alps.

Overall, 70% of the summits showed distinct increases in species richness. Their species pools have been enriched by invaders from below. This partially involved succession from a typical nival vegetation, dominated by nival specialists (e.g., *Androsace alpina, Poa laxa*), to a more subnival vegetation, even containing keyspecies of alpine grasslands (e.g., *Carex curvula*). At Piz Linard (3411 m), a mountain which has been monitored for 150 years in order to detect climate change impacts, vascular plant populations increased remarkably during recent decades (Braun-Blanquet, 1957, and personal observations). The rates of upward movement proved to be highly related to the geomorphological situation. Mountains consisting of solid rocks showed the highest increases in species richness.

The results underlined the importance of permanent plot research for detecting climate effects on vegetation patterns and, additionally, that measuring the sensitive reaction of high mountain plants requires observations at very fine spatial scales in relation to the microrelief. Therefore we established a net of observation plots on Schrankogel (3497 m), Tyrol, Austria, consisting of about 1000 permanent plots of 1 m^2 each, distributed over a topographically well-structured slope system. This transect study covers the transition zone from closed alpine swards to the open nival vegetation, at an altitude of about 3000 m., i.e., the alpine-nival ecotone. At the fine spatial scale of 1 m, migration effects should be detectable in a time interval of ten years.

A numerical analysis based on the permanent plot data revealed distinct vegetation types and ecological species groups in relation to the snow patterns and habitat stability. This provides evidence that species close to the altitudinal limits of plant life are not distributed by chance: they follow distinct ecological gradients.

Alternatively, we established about 70 similar observation plots at the high latitude limits of plant life, i.e., in the polar desert of Franz Josef Land (FJL), beyond 81° N. The two data sets show that the nival and the polar desert vegetation are comparable in terms of species richness and abundance of vascular plants.

Experiments

The FJL permanent plots are incorporated in the International Tundra Experiment (ITEX) network (Molau and Mølgaard, 1996). Franz Joseph Land is the northernmost field site of this circumpolar initiative, which attempts to perform climate change experiments in a standardised approach. A fully factorial treatment design will be applied to the FJL plots during the next seasons to simulate warming (using Open Top Chambers), enhanced nutrition, and enhanced availability of water.

These experiments will be extended to the Alps at the alpine and nival zone of Schrankogel during the next summer. Furthermore, transplantation experiments are planned to test the behaviour of selected species according to altered conditions of climate and competition.

Scenario modelling

Multivariate analyses of the ecotonal data set from Schrankogel revealed strong correlations between vegetation distribution patterns and the topographical and geomorphological environment. A fine-scaled Digital Terrain Model was used to obtain quantitative values of various geomorphological parameters, e.g., amount of scree, as well as of topographical parameters, e.g., relief roughness, distance to ridges, or exposure. Applying Canonical Correspondence Analysis (Ter Braak, 1986), environmental envelopes for about 20 vascular plant species could be defined. Their distribution patterns along the ecotone were predicted and displayed using a static, spatially explicit distribution model. Not only particular species distributions, but also those of plant assemblages as well as of ecological indices, e.g., species richness, can be displayed.

By assuming an altitudinal temperature gradient of –0.6°C per 100 m, and calculating distribution scenarios of past and predicted temperature regimes, it became obvious that upward migration of competitive species from the alpine zone will threaten the high biodiversity at the alpine-nival ecotone. The more or less homogeneous ecotonal belt of high species richness between the alpine and nival zone will disintegrate into small patches "trapped" in habitats with extreme terrain conditions. Such extreme habitats may be the only refuge for the typical species assemblages of today. A significant rearrangement of the high alpine vegetation must be expected, including many surprises.

These types of models will also be applied to the high arctic data sets. To link distribution predictions to directly measured climatic features, and to monitor their change, measurement networks using tiny temperature logger devices were established on FJL and on Schrankogel. This data will be used to improve the scenario modelling by introducing measured time-series of climatic factors on a fine spatial scale.

In the next step we plan to generate a measurement protocol for an indicator network of climate change effects in high mountain ecosystems. Two methodological approaches (Single Mountain- and Multi Summit-approach) will be carried out in the Alps to provide an example for the application in other high mountain systems. Pauli *et al.* (this volume) outline proposals to implement the indicator network on a global scale within the framework of the IGBP Mountain Workplan (Becker and Bugmann, 1997).

REFERENCES

Becker, A. and Bugmann, H. (eds.) (1997) Predicting Global Change Impacts on Mountain Hydrology and Ecology: Integrated Catchment Hydrology/Altitudinal Gradient Studies. IGBP. *IGBP Report 43*, IGBP Secretariat, Stockholm.

Braun, J. (1913) Die Vegetationsverhältnisse der Rätisch-Lepontischen Alpen. Neue Denkschr. Schweiz. *Naturforsch*. Ges., **48**. pp. 156–307.

Braun-Blanquet, J. (1957) Ein Jahrhundert Florenwandel am Piz Linard (3414 m). *Bull. Jard. Botan. Bruxelles*, Vol. Jubil. W. Robyns. pp. 221–232.

Gottfried, M. Pauli, H. and Grabherr, G. (1994) Die Alpen im "Treibhaus": Nachweise für das erwärmungsbedingte Höhersteigen der alpinen und nivalen Vegetation. *Jahrbuch des Vereins zum Schutz der Bergwelt*, **59**, München. pp. 13–27.

Grabherr, G. Gottfried, M. and Pauli, H. (1994) Climate effects on mountain plants. *Nature*, **369**: 448.

Grabherr, G. *et al* (1995) Patterns and current changes in alpine plant diversity. In. Chapin III, F. S. and Körner, C. (eds.). Arctic and Alpine Biodiversity: Patterns, Causes and Ecosystem Consequences, *Ecological Studies*, Vol. **113**, Springer Verlag, Berlin, Heidelberg. pp. 167–181.

Molau, U. and Mølgaard, P. (eds.) (1996) *ITEX Manual*. Second Edition. Danish Polar Center. Copenhagen.

Pauli, H. Gottfried, M. and Grabherr, G. (1996) Effects of climate change on mountain ecosystems – upward shifting of alpine plants. *World Resource Review*, **8**: (3): 382–390.

Reisigl, H. and Pitschmann, H. (1958) Obere Grenzen von Flora und Vegetation in der Nivalstufe der zentralen Ötztaler Alpen (Tirol). *Vegetatio*, **8**: 93–129.

Ter Braak, C.J.F. (1986) Canonical correspondence analysis: a new eigenvector for multivariate direct gradient analysis. *Ecology*, **67**: 1167–1179.

Brigitte Hanemann
Institut de Géographie Alpine, Grenoble, France and
Geographisches Institut der Technischen Universität München
Lucile-Grahn-Str. 46, 81675 München, Germany

The impacts of rock climbing on nature: a comparison of the management strategies adopted in France and Germany

Problem and methodology

The increasing number of rock-climbers in Germany and the expansion of their climbing sites have caused serious conflicts with the nature protection associations since the late 1980s. In some areas, this has resulted in a significant number of climbing bans. In France, the situation seems to be not so difficult for the climbers. A comparison between the environmental management strategies of the following French and German climbing sites analyses the perception of the impacts of rock climbing on nature in the two countries, to find out whether there exists a German and a French way of management or if the adopted strategies depend on the local situation and the local actors:

- France: Parc national des Ecrins, Parc naturel régional du Luberon, Gorges de l'Ardèche;
- Germany: Nationalpark Berchtesgaden, Oberes Donautal, Pegnitztal.

At every site, qualitative structured interviews with representatives of the local rock-climbing associations and representatives of the nature protection associations were undertaken and analysed.

General settings

The 800,000 French climbers share more than 1600 climbing sites below 1,600 m, whereas in Germany 80,000 climbers share 248 climbing sites. 60% of German climbers are affiliated to the German Alpine Club (DAV) or other climbing associations. In France, the affiliation is much lower.

The undeveloped natural climbing potential in Germany is very much lower than in France. It is presumed that in Germany this potential has been exhausted, whereas in France a great number of potentially accessible crags and cliffs remain. In addition, we must distinguish between alpine and other cliff areas.

Since 1986 in Germany, the *Biotopschutzparagraph* (§20c BNatG) has been in effect. This law globally protects a variety of biotopes, including rock ecosystems, from "serious" impact. The implementation of this federal law in the laws of the states varies very widely between legality of rock-climbing practices in Bavaria and Saxony to the illegality of rock-climbing in Baden-Württemberg and Rhineland-Palatinate. Climbing regulation depends on the state: in North Rhine-Westfalia, 90% of the crags are off-limits, while in Bavaria there are very few climbing bans. Since 1990, an organisational structure to supervise climbing sites representing the interests of rock-climbers and environmental protection has existed at local, regional and national levels.

In France, the *Arrêté préfectoral de protection de biotope* mainly concerns rock-climbing practices. This legal instrument is implemented by the Prefect to protect certain species and, in the case of protection of birds, it can include climbing bans. In 1994, 7.5% of the climbing sites were regulated or forbidden, but often the reason was not environmental protection but decisions of private owners. Some temporary *arrêtés de biotope* (bird protection) exist in the northeast of France, and numerous informal agreements in southern France.

Selected results

In Germany, there is regulation of the practice of rock-climbing: existing sites, used crags or routes are divided into zones, or are forbidden. The existing rock-climbing possibilities are reduced, as in subalpine ranges no new routes are permitted. In France, climbing sites are expanding: new climbing sites are discovered every year. Simultaneously, there is competition between climbers and nature protection associations: the latter try to protect the maximum of areas by *arrêté de biotope* for important species. Regulation or climbing bans in existing climbing sites are exceptional. In southern France, before a new route is created, climbers and ornithologists often decide about zones for birds to be kept free of bolts and routes.

In Germany, protection of flora and protection of fauna (falcons and owls) is determinant for the ecological management of climbing sites. For instance, "gardening" of routes has not been practiced for a long time. In France, ecological management chiefly concerns the protection of birds. "Gardening" is routine and the most recent manual "how to create new climbing sites" even explains the most efficient way of doing it. Gardening means safety.

Lowering-off-bolts are, in Germany, a contribution to environmental protection: they make summiting unnecessary. In France, lowering-off-bolts are common equipment at all sites: they are considered a contribution to climbing safety.

In France, climbing conventions are concluded between the FFME (Fédération Française de la Montagne et de l'Escalade) and the owner of the crag for reasons of liability because many crags are under private ownership or belong to the forestry service. In Germany 90% of crags are owned by the state and climbing conventions are concluded between the DAV, environmental associations and authorities to define the modalities for low-impact climbing.

In France, the climber is a guest of the owner of the rock. In Germany, the climber is guest of the rock biotope.

Conclusions

Generally, management approaches depend to a high degree on the local situation. However, the study shows that the management approaches adopted in France are characterised by an anthropocentric point of view, which means that they regulate the relationship between humans, whereas the management strategies adopted in Germany focus on the relationship between man and nature, which means that their main objective is to have as little impact as possible on the rock biotope.

Barry Meatyard
Environmental Sciences Research and Education Unit
Institute of Education, University of Warwick, Coventry CV4 7AL
United Kingdom

Koenigia islandica : a potential indicator of climate change in the UK

Koenigia islandica (Iceland Purslane) occurs widely in mountain and peri-glacial regions in the high latitudes. (Hultén, 1970). In Britain, *Koenigia* is known only on the islands of Mull and Skye along the west coast of Scotland (Rodwell, 1991), and is regarded as a relic of a flora that existed towards the end of the last glaciation (see Lusby *et al.*, 1996). This paper refers to an ongoing study of sites on the Ardmeanach Peninsula on the Isle of Mull where the distribution and abundance of *Koenigia* has been monitored since 1994. The Mull site is the most southerly in its present-day western European distribution (Jalas and Suominen, 1979), although it is recorded in fossil deposits from sites further south in Britain (Godwin, 1956). *Koenigia* is unusual as an arctic-subarctic species in being an annual and is thus held to be potentially sensitive to, and a possible indicator of, local climate change. At the limit of its geographical range on Mull, it is likely to be particularly vulnerable to such change.

The habitat of *Koenigia* on both Mull and Skye is highly specialised and comprises open basaltic gravel terraces at altitudes of 380 m to 500 m. The highest altitudes of its sites on Mull only just exceed the lower limit of this range. The gravel particles are from 0.5–5.0 cm in diameter and there is a semi-permanent flush of pH 6.6–6.8. Availability of such a specialised habitat is almost certainly a factor affecting its limited range in the UK. The distribution and abundance of *Koenigia* within the terraces has been recorded annually. The methodology has been to count individual plants in 1 m^2 quadrats. These have been placed randomly in specified areas within the terraces or have been fixed along transect lines. Each year, 64 quadrats were sampled of which 21 are fixed. Counts were made on or around the 30th July each year, when the plants are mature and in flower.

The results indicate that there has been a significant decline in the overall numbers of *Koenigia* plants since 1994. The fixed quadrats, in particular, show a reduction of up to 80% in some areas. The reasons for the decline are not clear, but competition and succession are unlikely as factors since there is no evidence of corresponding invasion of surrounding vegetation into the study sites. There is thus the possibility that climatic conditions are involved. Harrison (1997) has reviewed Scottish climate trends, and reports significant changes in recent years. In particular, there has been an increase in winter precipitation, a reduction in summer precipitation, and increased air temperature in the spring. Evidence also suggests that the west coast of Scotland has experienced a series of dry summers, with the early part of the year being particularly dry. In 1997, remote houses on Mull were experiencing water shortages. The summer of 1995 was exceptionally dry, with virtually no flush in the gravel terraces on Ardmeanach, and this is reflected in the overall reduction in abundance of *Koenigia*. This event has been calculated to occur with a 1-in-80 year frequency, but is predicted to occur three times in the 2050s (Cannell *et al.*, 1997).

A 30-year period is generally considered to be a minimum on which to base climate change evidence (Cannell *et al.*, 1997), and clearly this study currently does not meet this requirement. However, if there is a shift in the weather pattern which becomes part of longer-term climatic change, it is predicted that *Koenigia* will continue to show further decline in abundance and that it is an indicator of such change. Sætersdal and Birks (1997) have identified *Koenigia* as being one of a number of potentially sensitive species that are possibly vulnerable to climate change as a result of global warming in the mountains of Norway. Unlike the situation reported for vegetation in mountain areas in mainland Europe (Pauli, Gottfried and Grabherr, 1994) there is no scope for *Koenigia* to migrate upslope in Britain, since the terraces where it occurs on both Mull and Skye are at or near the summits of the hills concerned. There is therefore the possibility that the future existence of *Koenigia* in Britain is at risk as a result of climatic change.

REFERENCES

Cannell, M.G.R., Fowler, D. and Pitcairn, C.E.R. (1997) Climate change and pollutant impacts on Scottish vegetation. *Botanical Journal of Scotland* **49** (2) 301–313.

Godwin, H. (1956) *History of the British Flora*, p. 234. Cambridge University Press, Cambridge.

Grabherr, G., Gottfried, M. and Pauli H. (1994) Climate effects on mountain plants. *Nature* **369**, 448.

Harrison, S.L. (1997) Changes in Scottish climate. *Botanical Journal of Scotland* **49** (2) 287–300.

Hultén, E. (1970) *The Circumpolar Plants II*, Dicotyledons, p. 64. Almquist and Wiksell, Stockholm.

Jalas, J. and Suominen, J. (1979) *Atlas Florae Europaeae*, No. 4 Polygonaceae. Committee for Mapping the Flora of Europe, Helsinki.

Lusby, P. and Wright J. (1996) *Scottish Wild Plants*, p. 49–51. RBGE. The Stationery Office, Edinburgh.

Rodwell, J.S. (ed) (1991) *British Plant Communities*, Vol. 2, Mires and Heaths, p. 329–330. Cambridge University Press, Cambridge.

Sætersdal, M. and Birks, H.J.B. (1997) A comparative study of Norwegian mountain plants in relation to possible future climate change. *Journal of Biogeography* **24** 127–152.

Cathy Mordaunt
University of Stirling, Department of Environmental Science
Climate Laboratory, Stirling, FX9 4LA
United Kingdom

Association between weather conditions, snow-lie and snowbed vegetation in the Scottish Highlands

Snowbed vegetation contains both vascular plants and bryophytes. The latest snowbeds cover areas that have predominantly, if not exclusively, bryophyte flora; vascular plants are generally confined to the periphery of these snowbeds. It is hypothesised that the exclusion of vascular flora from the snowbed core is the result of the shortened growing season generated by late-lying snow, which the bryophyte flora are better able to tolerate. The snowbed bryophytes cannot, however, tolerate the competition offered by the vascular flora in the peripheral areas from which they are absent.

Data indicate that some bryophyte snowbed species (*Pohlia ludwigii* in particular) inhabit optimal conditions in the snowbed core, rather than tolerating sub-optimal conditions. Adaptation and acclimation responses, and reduction in performance with greater snow-lie duration observed in peripheral vascular species, indicate that these are inhabiting sub-optimal conditions in the snowbed periphery and cannot tolerate the stress of the reduced growing season in the snowbed core. Optimal thermal conditions are observed to occur during May and June. Late snow-lie forces the vegetation beneath to initiate and complete its annual cycle in sub-optimal thermal conditions. Bryophytes are less restricted by low temperatures (Furness and Grime, 1982) and demonstrate an ability to continue growing in near-freezing temperatures, through to October or November, when they are once more buried.

The relationship between snow-lie and climate is examined, using the hypothesis that snowbed loyalty in the Scottish Highlands is high, while duration of snow cover is variable. Snow-lie loyalty is the product of prevailing wind conditions, which are persistent and consistent in Scotland, leading to consistency in late snowbed location. However, the occurrence of mid-winter thaws at all altitudes increases variability in the duration of snow cover through accumulated snow depth. Climate, as measured by the Aonach Mor Automatic Weather Station, shows a year dominated by two distinct seasons, with (during 1992/3) a winter season of almost exclusively zonal flow, and a summer season with much more variable wind direction.

The evolution of Scottish snowpacks, with mid-winter thaws common (Green, 1975), generates snow of remarkably high density (Ferguson, 1984), which takes longer to melt for a given depth, and with accumulated depths of 20 m or more possible in certain locations. The steep altitudinal gradients of the Scottish mountains and the evolution of dense snowpacks generate much later-lying snowbeds than would be expected for mountains of this altitude and latitude.

Snow-lie in the Highlands is decreasing at all altitudes, according to the Snow Survey of Great Britain. An increase in zonal flow in winter and spring can be seen from the pressure difference between Cape Wrath and Eskdalemuir, which promotes loyalty to location in late snowbeds and ensures deep accumulation from snow- and wind-drift. Furthermore, with increased zonal flow, the steeper lapse rates commonly found in polar maritime air masses may lead to an increase in snowfall at higher altitudes and increases in accumulation, generating a steepening of the snow-lie gradient with altitude and promoting the survival of snowbed vegetation.

REFERENCES

Ferguson, R.I. (1984) Magnitude and modelling of snowmelt runoff in the Cairngorm mountains, Scotland. *Hydrological Sciences Journal*, **29**: 49–62.

Furness, S.B. and Grime, J.P. (1982) Growth rate and temperature responses in bryophytes. II. A comparative study of species of contrasted ecology. *Journal of Ecology*, **70**: 525–536.

Green, F.H.W. (1975) The transient snow-lie of the Scottish Highlands. *Weather*, **30**: 226–235.

Simon Patrick and R. Flower
Environmental Change Research Centre, University College London
26 Bedford Way, London WC1H 0AP
United Kingdom

Measuring and modelling the dynamic response of remote mountain lake ecosystems to environmental change: an introduction to the MOLAR project

The arctic and alpine regions of Europe represent the most remote and least disturbed environments in Europe, yet they are threatened by acid deposition, toxic air pollutants and climate change. The remote lakes that occur throughout these regions are especially sensitive to these threats for a number of related reasons:

- many have little ability to neutralise acidity because of their low base status;
- nitrate levels are higher because their catchments have little soil and vegetation to take up nitrogen deposition;
- toxic trace metals and trace organics accumulate in the food chain more easily, and some pollutants (e.g., mercury, volatile organics) accumulate preferentially in cold regions;
- climatic warming in Europe is predicted to be greatest in arctic and alpine regions.

Because of this sensitivity, remote mountain lakes are not only vulnerable to environmental change, but also excellent sensors of change. Their high-quality sediment records can be used to infer the speed, direction and biological impact of changing air quality and climate.

The MOLAR (Monitoring of Lakes in Arctic and Alpine Regions) project builds on the success of previous projects funded by the European Commission (EC): AL:PE.1 (Acidification of Remote Mountain Lakes: Palaeolimnology and Ecology) and AL:PE.2 (Remote Mountain Lakes as Indicators of Air Pollution and Climate Change), which represented the first comprehensive study of remote mountain lakes at a European level.

The AL:PE projects assessed the status of remote mountain lakes throughout Europe on the basis of sediment core records and chemical and biological surveys. The MOLAR project focuses on detailed studies of a smaller number of key sites to provide high-resolution data on their temporal dynamics that can then be used to develop and calibrate predictive models.

The MOLAR project has four overall objectives:

- to measure and model the dynamic responses of remote mountain lake ecosystems to acid (sulphur plus nitrogen) deposition;
- to quantify and model pollutant (trace metals, trace organics) fluxes and pathways in remote mountain lakes and their uptake by fish;
- to measure and model the temporal responses of remote mountain lake ecosystems to climate variability on seasonal, inter-annual and decadal time-scales.
- to continue the development of a high-quality environmental database on remote mountain lake ecosystems in Europe and to disseminate results widely to enhance public awareness, environmental education and environmental decision making.

The main deliverable generated by the project will be predictive models for acidity, pollutant flux and climate variability that can be used in scenario assessment studies, especially those scenarios associated with present and forth-coming protocols of the United Nations Economic Commission for Europe (UNECE) and General Circulation Model (GCM) predictions for Europe. A desirable future objective would be the linking of these models to evaluate the interaction between acidity, trace pollutants and climate.

In addition to model development, much of the field and laboratory work proposed is innovative for studies of such remote sites, especially:

- the focus on the seasonal dynamics of the lake systems;
- the emphasis on nitrogen deposition and its biological impact;
- the study of microbial food webs in relation to acidity;
- the on-site collection and measurement of atmospheric pollutants;
- the use of radio-tracers to validate pollutant transport models;
- the study of trace metals (especially mercury) and trace organic uptake by fish;
- the on-site monitoring of climatic conditions and their relationship to water column behaviour;
- the development of a methodology to infer climate trends from the high resolution analysis of recent sediments.

These objectives were unachievable four years ago because of the almost complete lack of information on remote arctic and alpine lakes. It is now possible to carry out such work because of the knowledge gained about individual sites from the AL:PE project with respect particularly to accessibility, morphometry, chemistry, biology, sediment accumulation rate and pollution status.

MOLAR is funded by the EC Environment and Climate Programme with assistance from the programme for international cooperation (INCO). It is coordinated by the Norwegian Institute for Water Research (NIVA) and the Environmental Change Research Centre (ECRC) at University College London, and involves the collaborative participation of 23 laboratories throughout Europe.

Therese Perrin-Sanchis
Cemagref, Domaine Universitaire – BP76
38402 St Martin d'Heres
France

Alpine research: a new dynamic to develop multi-disciplinary scientific cooperation across borders

The 'Agenda 21' action programme set by the Rio Conference for sustainable development in mountain regions (Chapter 13); the fifth European Union Environmental Action Programme; the fifth European Research and Development Programme currently being prepared; and the Alpine Convention and its protocols, all constitute major elements driving new efforts to develop multi-disciplinary scientific cooperation across borders.

Given the above major orientations, mountain environments will continue to be needed in their widely-recognised role as "Idea Laboratories". Mountain communities have long been the creators of territorial systems based on an ever-fragile balance between the necessities of economic development and of those concerning the protection and preservation of a particularly sensitive environment. Mountain research has therefore always been particularly interested in progress made by pooling knowledge and multi-disciplinary approaches, in order to analyse and understand social behaviour and the interrelations between people and their environment, as well as the underlying processes of environmental change. Alpine research, therefore, represents an array of competencies and coordination capabilities that put it in a particularly favourable position to develop original solutions with regard to new forms of regional development.

Alpine research organisations have undertaken to develop their work along the above lines. A special showcase for this approach is the Alpine Forum, which was held for the first time in Disentis, Switzerland, in 1994. The second event, in Chamonix, France, in 1996, focused on the new conditions governing development of the Alpine region. The meeting confirmed the desire to provide general problems with solutions contributing to general well-being, and applicable at the regional level. This work is carried out in conjunction with that of the "Alpine Information and Observation System", initiated by the Alpine Convention. The next Alpine Forum, in Garmisch-Partenkirchen, Germany, in 1998, will focus on the forces of change in the Alps.

Originally constituted for the needs of the Alpine Forum, the International Scientific Committee for Alpine Research has devoted itself, in addition to permanent missions, to motivating dialogue and exchange concerning alpine research; encouraging interdisciplinary and scientific collaboration across borders; developing relevant questions concerning the Alpine Arc by international research programmes; evaluating and studying research needs of the Alpine Convention; and counselling its task forces or committees concerning research. For this, it has 16 continuous members from different disciplines, in natural, social, and human sciences, who represent six countries and the European Commission (EC).

Alpine research: Perspectives and priorities

A study is currently underway, with support from EC-Directorate-General XII, in relationship to the 5th European Research and Development Programme which is currently being prepared. The study attempts to determine and formalise the presentation of research subjects that are relevant in terms of perspectives offered by Alpine research, by undertaking to further the analysis of the current situation initiated at the Alpine Forums. The results of this work may serve as common reference material not only for researchers in their efforts to cooperate and exchange information, but also for actors and applicants in setting up projects. With regard to multi-disciplinarity and analysis of the interrelationships between people and their environment, mentioned above, the Forums contributed to the emergence of a central topic: 'Systems implemented in use of territory and the forces behind change in the Alps: Prospective analysis of territorial restructuring modes, and environmental forces at work'. This is currently broken down into four major aspects:

- diversity and dynamics of alpine zones;
- territorial restructuring process;
- stabilisation and control of environmental dynamics;
- social, cultural and political appropriation of dynamics.

The current structure used to express a central topic and supportive questions is still a simple outline that must be confirmed and positioned in terms of the perspectives for Alpine research. The study will consequently go into five main aspects:

- identification of relevant themes for Alpine research;
- structuring of themes and scientific subjects;
- enhanced expression of social requests; positioning of scientific themes with respect to the set of relevant subjects concerning the development and the protection of the Alps, as expressed in particular in the Alpine Convention and its protocols;
- positioning of Alpine research, in terms of both what makes it specific and similar in its themes, with respect to other territories, whether in mountain areas or not;
- positioning of the means allocated to these subjects, as well as of cross-disciplinary and cross-border exchanges and collaboration.

Five institutions are collaborating in realising this study:

- Cemagref Grenoble, division Développement des territoires montagnards (Cemagref-DTM), France (coordinator);

- Université Joseph Fourier, Grenoble, Institut de Géographie Alpine (IGA), France;
- Académie Suisse des Sciences Naturelles (ASSN) – task force for alpine research, Switzerland;
- Alpenforschunginstitut, Garmisch-Partenkirchen (AFI), Germany;
- Europäische Akademie Bozen, Fachbereich "Alpine Umwelt" (EAB), Italy.

The two French organisations belong to the "Pôle Universitaire et Scientifique Européen de Grenoble"; Austrian and Slovenian collaborators are also involved.

Bill Slee, Helen Farr and Patrick Snowdon
Department of Agriculture, University of Aberdeen
581 King Street, Aberdeen AB24 5UA
United Kingdom

The economic impact of different styles of rural tourism: examples from France, Portugal and the UK

The tourism sector has been a major growth sector in many rural areas of developed countries in the last two decades, at a time when there has been a major decline in the relative contribution of agriculture to rural well-being. Tourism may appear to be a particularly attractive option in remote and disadvantaged areas, where land quality for agriculture may be low, but whose environmental qualities, as a result of traditional and low-intensity (and often low-profitability) agricultural activity, may be very high.

Although there may be some scope for improving farm household incomes through developing the farm sector of these disadvantaged regions, the greatest scope for enhancing farm household well-being is likely to lie in off-farm work or diversification into non-farm enterprises. In the UK, farm tourism is the most common form of on-farm diversification (McInerney and Turner, 1989; Dalton and Wilson, 1989), and tourism is the single largest employer in many rural regions in the north and west of Britain.

The role of tourism in rural, national and supranational economic restructuring has been widely stressed (HMSO, 1989; Grolleau, 1988), and tourism provides a classic example of a post-productionist product from the changing rural economy. The European Union (EU) has actively supported rural tourism since the late 1980s through spending from the reformed Structural Funds and the LEADER Community Initiative. In the first round of LEADER spending, over 40% of all support went to tourism projects.

There has been widespread advocacy of soft/green/sustainable tourism from the environmental lobby and by community development activists. The principal state and voluntary environmental organisations in the UK have supported the idea of green tourism (RDC, ETB and CC, 1992). However, despite this popular acclaim, there has been no structured economic analysis of the merits of soft/green tourism in comparison with more conventional approaches to tourist activity.

This study is based on the detailed investigation into rural tourism in six economically disadvantaged regions in three countries: France, Portugal and the UK. An *a priori* distinction was made between hard and soft tourism: the former comprising the hotel sector and other large-scale holiday developments; and the latter, tourism related to the land-based sector. The study areas were selected so as to include areas with hard and soft tourism components in different proportions.

Major differences were found both between and within countries in the conceptualisation of rural tourism and the role of rural tourism in rural development. These differing perceptions inform institutional actions and shape pathways of tourism development, sometimes giving tourism close connections with the land-based sector and elsewhere clearly separating tourism from the land.

A model of the tourist sector of the local and regional economies, based on partial input–output analysis, was constructed. This yielded a wide range of results relating to the overall economic impacts of spending of soft and hard tourists, as defined by accommodation sectors used. In this study, the proportional multiplier method was used, which estimates the direct, indirect and induced impact per unit visitor spend, rather than as a ratio of direct, indirect and induced to direct impact. In four out of six areas, the soft tourist sector generated higher levels of local impact per unit tourist spend (Table 1), and in five out of six areas, soft tourism generated a higher level of employment (Table 2).

Table 1: Proportional multipliers for hard and soft tourism in the core areas and the core + extended areas

Region	Core area		Core + extended area	
	Hard tourism	Soft tourism	Hard tourism	Soft tourism
Livradois-Forez (France)	0.31	0.55	0.43	0.60
Vercors (France)	0.30	0.37	0.35	0.40
Douro (Portugal)	0.46	0.40	0.52	0.47
Peneda Geres (Portugal)	0.45	0.46	0.57	0.59
Badenoch and Strathspey (UK)	0.22	0.30	0.30	0.42
Exmoor (UK)	0.23	0.21	0.34	0.34

Table 2: Standardised employment created in core areas per 100,000 ECU of visitor spending in soft and hard tourism

Region	Direct standardised jobs		Total standardised jobs	
	Soft tourism	Hard tourism	Soft tourism	Hard tourism
Livradois-Forez (France)	5.6	1.5	6.3	1.9
Vercors (France)	2.8	1.7	3.1	2.0
Douro (Portugal)	9.2	3.7	10.3	4.8
Peneda Geres (Portugal)	10.2	4.7	12.4	6.7
Badenoch and Strathspey (UK)	4.0	2.2	4.8	2.8
Exmoor (UK)	2.3	2.1	2.7	2,6

There were significant differences in the relative performance of the soft and hard sectors which can be attributed to the particular types/styles of soft and hard tourism in the different study areas. It should, however, be noted that hard tourism in large hotels and holiday villages generally generates a higher *per diem* spend than in the soft sector where farm bed-and-breakfast, self-catering and camping are the main forms of tourist activity. For example, where soft tourism was dominated by camping, much lower levels of employment creation resulted. Generally, but not decisively, soft tourism strategies will generate more scope for local benefits per unit spend by tourists.

The institutions which support rural tourism are varied, and this range of institutions has resulted in national, regional and local differences in the implementation of support strategies. There is a distinct divergence between the tourism-centred agencies and those centred on rural development with regard to their perceptions and interventions in relation to rural tourism. EU involvement further complicates an already complex set of interventions. To date, the principal forms of EU support have been through the Structural Funds and community initiatives such as LEADER and LIFE, thus giving a strongly rural development character to these interventions.

In local rural economies, with a limited network of business suppliers, it is inevitable that the greatest economic impact will arise from the direct effect. The scale of indirect and induced benefits is conditioned by the characteristics of the local economy and the extent to which the tourism sector, consciously or unconsciously, is linked to the local economy. It is possible to enhance the multiplier by differentiating the tourist product through its use of locally available foods and drink and its identification with local craft and artisanal products. Private, public or partnership action to increase these linkages is likely to enhance the local multiplier.

Agro-tourism clearly has the capacity to offer significant benefits to households and to the wider economy in disadvantaged rural regions. It can be seen both as a household adjustment strategy, and as part of a wider regional economic strategy. The emergence of wide ranging, locally-based development projects, such as LEADER, offers additional scope for the development of stronger regional linkages and synergies in local economic development. However, for this synergy to be realised, local institutions and external bodies must increasingly form partnerships in which it is essential that all interests are represented, if unbalanced and unsuccessful developments are to be avoided.

REFERENCES

Dalton G. and Wilson, C. (1989) Farm Diversification in Scotland. Scottish Agricultural College: Aberdeen.

Grolleau, E. (1988). Rural Heritage and Tourism in the EEC. Commission of the European Communities: Brussels.

Her Majesty's Stationery Office (HMSO) (1989) Pleasure, Leisure and Jobs, the Business of Tourism. HMSO: London.

McInerney, J., Turner, M. and Hollingham, M. (1989) Diversification in the Use of Farm Resources. Agricultural Economics Unit: Exeter.

Rural Development Commission, English Tourist Board and Countryside Commission (RDC/ETB/CC) (1992) Green Light: a Guide to Sustainable Tourism Development. RDC, Salisbury.

Ian Thompson
Department of Geography & Topographic Science
University of Glasgow, Glasgow G12 8QQ
United Kingdom

Sustainable rural development in the high French Alps: evidence from the Parc National de la Vanoise

The concept of 'sustainable' development enjoys considerable vogue, but a precise decision is difficult to achieve beyond the notion that development should not compromise the environmental legacy and physical support systems left for future generations. This objective is clarified when it is disaggregated into four interacting components; ecological, economic, cultural and political. The ecological dimension requires that the fundamental environmental systems are maintained in a state of equilibrium. The economic dimension contains more uncertainty but implies that development should be compatible with the maintenance of future production and consumption, employment opportunities and an adequate material standard of living. Cultural sustainability implies the continuity of a distinctive heritage and its infrastructure of vernacular buildings, field patterns and archaeological sites, together with the encouragement of traditional customs and folklore. Finally, sustainability must repose on a robust political system and framework of legislation, from the macro scale to local by-laws, which is capable of mediating conflicts.

The four components of sustainability are in constant interaction and often in conflict. Their interplay is more easily decipherable at a local scale and especially within the context of fragile environments subject to development pressures. Accordingly, the proposed four-fold classification has been applied in a pilot study to a high mountain commune in the French Alps, Peisey-Nancroix, situated within the Parc National de la Vanoise, (Thompson, 1997). The commune was selected as being representative of the traditional pastoral transhumance economy of the Tarentaise region, which has a modern history of depopulation and a recent revival based largely on tourism.

Peisey-Nancroix commune is a left-bank hanging tributary of the Isère valley. The valley descends from glacier-bearing summits, rising to 3,417 m, and has the characteristic layering of land resources and settlement. Improved land is situated in the valley floor at its lowest elevation. Above the alp bench, seasonal grazing is occupied by a number of temporary settlements, the 'montagnettes', until recently occupied by farmers during the seasonal movement of cattle on their way to the 'alpages', the extensive communal summer grazing. The successive use of the resource layers was integrated via transhumance and reposed on considerable local knowledge of environmental conditions underpinned by a stable society and cultural life.

The traditional pastoral economy, although sustainable in ecological terms, had a relatively fixed labour demand, and population growth could only be sustained by exploiting other resources. Initially this concerned silver-lead mining which allowed the commune to triple its population to a peak of 1630 in 1837. With the closure of the mine in 1865, the commune was reduced to its pastoral basis, and by 1936 had reached its nadir of 417 inhabitants. Since that date, Peisey-Nancroix has experienced a slow recovery, almost entirely related to tourism, attaining 521 inhabitants in the 1990 census. On the trends of the last three decades, the commune appears to have achieved a sustainable equilibrium. This hypothesis may be tested against the four criteria enunciated above.

In political terms, two politico-legal circumstances impinge directly on the commune. The Parc National de la Vanoise was inaugurated in 1963 and includes the entire commune. A distinction is drawn between the totally protected core zone and the peripheral zone which functions as a reception area. Secondly, although development is permitted in the peripheral zone, it is subject to strict controls within a land-use plan. In principle, a political and planning framework exists with an implicit function of sustainability. Most significantly, the designation of the core area has excluded extensive downhill skiing which might otherwise have triggered large-scale resort development.

Cultural sustainability is more intangible. In practice, it is easier to conserve physical objects and vistas than to engineer social attributes. With financial assistance from the Park Authority, the valley's building heritage has been made the subject of an inventory with the specific aim of conservation. Cultural groups have proliferated and monuments have been restored, reflecting both local sentiment and recognition of a tourist asset. A strict implementation of the land-use plan is essential to protect the quality of the built environment, especially in view of the proliferation of second homes and the haphazard construction of a small scale ski-resort at Plan Peisey which pre-dates the present planning controls.

In economic terms, pastoralism and tourism still exist and reinforce each other. The farming economy and transhumance persist, but on the basis of fewer, larger operations. The smaller holdings commonly depend on a second source of employment, invariably in tourism. The dominant agricultural commodity, local cheese, finds a ready market. Whereas agricultural employment has contracted, tourism displays the reverse trend. The emphasis on informal tourism with a light infrastructure has had a limited environmental impact so far.

Ecological sustainability has been maintained by the combined effects of the National Park legislation, local planning and the environmentally-adjusted pastoralism. Apart from natural hazards, the main problems concern the disposal of waste water and the threat of further ski lift development to link with neighbouring resorts less constrained by Park regulations (Les Arcs, La Plagne).

This pilot study indicated that, even in fragile environments, sustainable development can be a realistic objective provided certain critical thresholds are not crossed. It is also a truism that the local population tends to be the best guardian of environmental quality, provided that it is underpinned by appropriate and effective legislation.

REFERENCE

Thompson, I.B. (1997) Sustainable Development in the High French Alps, *Scottish Association of Geography Teachers Journal*, **26**, 59–67.

C.E. Thorn[1], R.G. Darmody[2], J.C. Dixon[3], J.P.L. Schlyter[4] and C.E. Allen[1]
[1]Department of Geography, University of Illinois at Urbana-Champaign
Urbana IL 61801, USA
[2]Department of Natural Resources and Environmental Sciences, University of Illinois
[3]Department of Geography, University of Arkansas
[4]Department of Geology, University of Lund

Chemical weathering in the Scandinavian Mountains

Geomorphologists have long operated under the assumption that mechanical weathering processes dominate in cold environments, and that chemical weathering processes are absolutely and relatively minimal. This perspective stems from: (1) the historically entrenched, but poorly established, concept that freeze-thaw weathering is powerful and ubiquitous in cold environments; (2) the view that naturally occurring cold temperatures greatly reduce, if not eliminate, chemical reactions; (3) the view that cold regions are arid. Integration of the preceding ideas has resulted in far too little research into the nature and magnitude of chemical weathering processes in arctic and alpine environments by periglacial geomorphologists.

While the efficacy of freeze-thaw processes is widely assumed by periglacial geomorphologists, the underpinning theory and laboratory experimentation vindicating this view are sketchy or equivocal. Even if the freeze-thaw process is important, this does not mean that chemical processes are either negligible or unimportant. An important intellectual step for periglacial geomorphologists is to abandon their pre-occupation with macro- and meso-scale air climates and to focus upon micro-scale ground climates where chemical weathering actually takes place (Pope et al., 1995). In reality, thin films of water are commonly available in most arctic and alpine ground regimes (i.e., they are not deserts in the relevant sense); nor do naturally occurring temperatures significantly inhibit (let alone eliminate) chemical reactions (Tamm, 1924). Consequently, the nature, relative and absolute importance of chemical processes in arctic, alpine, and arctic-alpine environments should be regarded as important open questions meriting immediate attention.

Rapp (1960) established that in Kärkevagge, a valley in arctic Sweden, stream solutes constituted some 48% of the material being removed, and also that chemical weathering was the single largest source of losses. Work presently being written up (e.g., Dixon et al., 1995), and stemming from detailed analyses of chemical processes undertaken in the early 1990s, shows that chemical processes typical of "temperate" environments are active and exhibit spatial variability within the valley. Investigations of chemical processes include: examination of rock crusts, rinds, skins, and varnishes using electron microscopy and microprobe; measurement of spatial variability of chemical weathering within the valley, using machine-polished disks and machine-tumbled granules of granite, dolomite, and limestone; detailed pedogenic studies and soil mapping; and a brief period of water quality analysis.

Matthews (e.g., 1992) has created a detailed picture of the ecological trends on a glacial foreland following glacial retreat at Storbreen, Jotunheimen Mountains, Norway. As part of his more than 20-year study, Matthews has precisely dated a large number of surfaces on the foreland, including many moraines. Using Matthews's work as a temporal framework, Darmody *et al.* (1987) and Darmody and Thorn (1998) have shown that clay mineralogy (taken as an indicator of chemical weathering) progresses quite rapidly along lines similar to those found in "temperate" environments. Discernible differences in mineralogy were detected along a transect of constant elevation but differing ages, as well as along a moraine of constant age but varying elevation. These differences have developed since 1750 A.D., and while the differences are fairly subtle, they reflect surprisingly rapid development in a regime that traditionally would be assigned a relatively inactive chemical regime.

In summary, chemical processes of a nature found in "temperate" environments appear to be active in the arctic/alpine mountains of Scandinavia. They appear to be temporally and spatially quite sensitive and may make useful indicators of environmental change as changes in process rate and/or nature must precede change in landform or landscape (landform assemblage). Because chemical processes are generally "saprolitic", i.e., they represent loss of mass, but leave a skeletal residuum, they are readily overlooked by geomorphologists relying on pure human perception. This explains much of the lack of attention paid to chemical processes in alpine and/or arctic environments.

REFERENCES

Darmody, R.G., Thorn, C.E. and Rissing, J.M. (1987) Chemical weathering of fine debris from a series of Holocene moraines: Storbreen, Jotunheimen, southern Norway. *Geografiska Annaler*, **69A**: 405–413.

Darmody, R.G. and Thorn, C.E. (1998) Elevation, age, soil development, and chemical weathering at Storbreen, Jotunheimen, Norway. *Geografiska Annaler*. (In Press)

Dixon, J. C., Darmody, R. G., Schlyter, P. and Thorn, C. E. (1995) Preliminary investigation of geochemical process responses to potential environmental change in Kärkevagge, northern Scandinavia. *Geografiska Annaler*, **77A**: 259–267.

Matthews, J.A. (1992) *The Ecology of Recently-Deglaciated Terrain.* Cambridge University Press, Cambridge.

Pope, G. A., Dorn, R. I. and Dixon, J. C. (1995) A new conceptual model for understanding geographical variations in weathering. *Annals of the Association of American Geographers*, **85**: 38–64.

Rapp, A. (1960) Recent development of mountain slopes in Kärkevagge and surroundings, Northern Scandinavia. *Geografiska Annaler* **42**: 71–200.

Tamm, O. (1924) Experimental studies on chemical processes in the formation of glacial clay. *Sveriges Geologiska Undersökning Årsbok*, 18.

Jerry Thomas Warren [1] and Ivar Mysterud [2]

[1]High Mountain Ecological Research Station – Finse
c/o Biological Institute, Box 1050, Univ. of Oslo, 0316, Oslo, Norway
[2]Biological Institute, Box 1050, Univ. of Oslo, 0316 Oslo, Norway

Conflicting interests on Hardangervidda (Norway): traditional resource use in a modern setting

The Hardanger mountain plateau (Hardangervidda) is an area of alpine tundra located in south-central Norway, covering an area of nearly 10,000 km^2. For at least 5000 years, wild and domestic ruminants have grazed on Hardangervidda, and the long history of grazing has directly or indirectly influenced nearly all plant and animal communities on the plateau (Wielgolaski, 1997). The conversion of Hardangervidda's endemic vegetation by reindeer, cows, goats and sheep into meat, milk and skins is the oldest and historically most important use of the plateau's natural resources (Warren and Mysterud, 1995). Competition between grazing and hunting interests in utilising these resources has traditionally been intense, and laws promoting and regulating grazing on Hardangervidda date back to the Middle Ages.

Reindeer hunting and sheep grazing, in more modern and regulated forms, are still economically and culturally important activities on and around much of Hardangervidda, though the relative importance of these activities varies locally. That is to say that the costs and benefits of hunting and grazing are by no means evenly distributed around Hardangervidda. Subsequently, conflicts between grazing and hunting interests are still common. Conflicts between local and regional sheep farmers on the western part of the plateau have also arisen. Larger flocks from coastal areas have for decades leased grazing allotments on the plateau commons. While only able to maintain much smaller flocks due to limited winter forage, many local intermountain sheep farmers (often allied with local hunting interests and cabin owners) disfavour the use of ("their") common grazing resources by outsiders.

Recently, these conflicts have, on Hardangervidda and in other mountain areas, spawned ecological investigations into possible resource competition or other negative interactions between domestic sheep and wild reindeer (Bergmann, 1997; Gjerde, 1998). Such investigations are no easy task, however, due to the extensive habits of both reindeer and sheep. Current patterns of resource use, especially by reindeer, may also be the result of historical circumstances that can no longer be observed.

During the last century, use of Hardangervidda, as well as most other mountain areas in Norway, has changed dramatically. The development of hydroelectric facilities, for example, has fragmented reindeer habitat. Two winter-open roads and a railway that now transverse Hardangervidda have led to further fragmentation. Road and railway construction has also made once relatively remote areas more accessible for hunters, sheep farmers and especially tourists (Warren and Mysterud, 1995). Increases in national wealth, and subsequently disposable personal income and leisure time, have also prompted a dramatic increase in the number of hikers, bikers and other vacationers on and around Hardangervidda, a trend that will likely continue. A large portion of the central plateau is now a national park. Marked trails, cabins and hotels are numerous (the latter of which lie outside the park), and construction continues in areas adjacent to the park. Consequently, circumstances surrounding negative interactions between domestic sheep and wild reindeer (and between sheep owners and reindeer hunters), both actual and construed, are seldom clear and must be viewed in light of increasing human activity and newer forms of mountain resource use. Ecology-based field investigations of competition, resource partitioning, activity budgets etc. are difficult at best, and such studies alone are certainly inadequate.

Until recently, local governments around Hardangervidda have played an important role in the policy-making and management process. Municipal governments, however, have proved in many (most) cases to be partial to either tourism, hunting and/or sheep grazing interests, or some combination of these, again depending upon the relative importance of the activities to the local economy. This has led to inconsistent policies with respect to the regulation of tourism, grazing, development and the use of motorized vehicles, such as snowmobiles. The result has been a poorly coordinated effort to manage Hardangervidda and its resources. Subsequently, management stands now to be placed to a greater extent under central control. It remains uncertain, however, whether central authorities will be able to strike a politically and ecologically sound balance between traditional hunting/grazing interests and the so-called non-consumptive provision of amenities.

In summary, we are faced with two challenges: 1) The ecologist and resource manager must distinguish between actual animal-animal interactions (e.g., interspecific competition) which may have negative effects on the wild reindeer population, and those which are negatively construed by a growing number of competing interest groups. 2) Local authorities must either exert a more coordinated effort (i.e., be willing to look beyond purely local interests) in managing natural resources of national importance, or relinquish management responsibility to central authorities.

REFERENCES

Bergmann, C. (1997) Wild reindeer and rygja sheep's use of summer range in Setesdal-Veihei and Ryfylkeheiene. M.S. Thesis. Institute of Animal Husbandry, Norwegian Agricultural University, Ås. [in Norwegian]

Gjerde, G. (1998) Effects of sheep grazing in Setesdal- and Ryfylke wild reindeer district. M.S. Thesis. Department of Botany, Norwegian University of Science and Technology, Trondheim. [in Norwegian]

Warren, J.T. and Mysterud, I. (1995) Sheep, wild reindeer and resource use on Hardangervidda past and present. Biological Institute, University of Oslo. [in Norwegian]

Wielgolaski, F.E. (ed.) (1997) Polar and alpine tundra. *Ecosystems of the World 3*. Elsevier, Amsterdam.

Wolfgang Weinmeister
Institute for Avalanche Control, University of Agricultural Science
1190 Vienna
Austria

Sustainable mountain risk engineering

Our mountain regions are endangered by numerous natural hazards. The natural hazards which occur most frequently in the European Alps and cause considerable damage are unleashed by precipitation (heavy rainfall, snowfall). Even snowmelt may be the reason. Landslides and rock falls may also be triggered by earthquakes. Some conditions of the complex interlinked natural disasters-society system influence the state of measures against hazardous risks. The budget used for the control of risks is restricted. The costs of damage increase faster than society is able to spend money for protection measures.

Studies of the costs of damage show that a significant proportion results from damage to technical control structures. These are becoming increasingly susceptible and have to be repaired. Appropriate maintenance measures to ensure a certain standard of security are indispensable, but are likely not to be realised in times of emergency and economic downswing. Inappropriately-constructed structures have a tendency to collapse. The current methods of construction for protection have high repair costs and high expenditures for maintenance. These will cause problems for financing when economic conditions are bad.

In Austria, 164.2×10^9 AS have been spent for risk mitigation since 1884. The yearly budget available for maintenance and new mitigation works is 1.5×10^9 AS. This is only 0.9 % per year of all investments until now which appears insufficient for adequate maintenance.

Examples of torrential floods in Austria, Switzerland or Spain (e.g., the big Arás disaster of 7 August 1996 in the Spanish Pyrenees: Garcia-Ruiz *et al.*, 1996) demonstrate that not every constructed safety precaution is able to prevent such disasters. Additionally, in some cases, the malfunction of the safety precaution increases the damage.

A study of the energy content of different control measures for torrent control indicates that protection structures, such as checkdams, made of concrete contain a lot of energy (Hitsch and Weinmeister, 1992) or need it for repair or maintenance. If we consider the increasing costs of energy in the future, the costs of maintenance will also rise. The expected economic downswing and the necessity of maintaining structures to avoid an increase in disasters will lead to a fatal financial bottleneck if we want to retain our present security level.

There are a number of criteria for sustainable development relating to disaster mitigation (Table 1):

- permanence of mitigation measures;
- low resource effort and resource consumption (especially energy);
- low economic and ecological total costs (construction expenses and of the maintenance costs, including disturbance of the ecological system and biodiversity);
- consideration of the natural geomorphologic and biological processes and the utilisation of them for the mitigation measures;
- consideration of system connections and interactions.

The restriction of the budgets have to lead to innovative security concepts with less technical and therefore energetic intervention.

Table 1: Criteria for sustainable disaster prevention

Disaster mitigation methods/criteria	Perman- ance of mitigation measures	Small resource effort and resource consump- tion	Small economic and ecological total costs	Consider- ation of the natural processes	Consider- ation of the system connect- ions
Avoiding endangered areas	+	+	+	+	+
Measures of land management, particularly watershed management, forest management and agro-foresting	+	+	+	+	+
Soil-bio-engineering works	+	+	+	+	+
Technical measures					
flood wall, dike	o	o	o	−	−
check dam	o	−	−	o	−
beadload-dosing dam, beadload-sorting dam	o	−	o	o	−

+: positive effect; o: intermediate effect; −: negative effect

Hazard zoning, warning-systems, alarm-systems and evacuation

Land management measures (watershed management, forest management and agro-forestry) are also critical because the vegetation-soil-complex has a buffer function on geomorphological processes. It can diminish runoff, stop erosion, restrain shallow landslides, slow down falling rocks in forests, and prevent the sliding of snow cover which can develop into an avalanche. To avoid critical technologies, such as certain road and ski run construction practices, is also important. Soil-bio-engineering works have a low energy content and the ability of self-regeneration.

Technical measures are the main elements of protection in developed countries. Austria, for example, has a tradition since 1884. But not all of construction measures correspond to the criteria of a sustainable development.

The 1993 Mississippi flood or the 1997 floods in Central Europe show us the limits of events and some mistakes with regard to extreme discharge if dikes are built. Thus, we can see that it is possible to develop a mitigation system that fits with sustainable development. The correspondence is dependent on the degree of development and the kind of measures.

REFERENCES

Garcia Ruiz, J.M., White, S., Marti., C., Valero, B., Errea, M.P. and Villar, A.G. (1996) La catástrofe del Barranco de Arás (Biescas, Pirineo Rangoon's) y su contexto espacio-temporal. Instituto Pirenaico de Ecología, Zaragoza.

Hitsch, R. and Weinmeister, H. W. (1992) The flow of energy by realisation of different methods of construction in torrent control. *Interpraevent 1992* – Bern, vol. 4, pp. 279–290.

Georg Wiesinger
Federal Institute for Less-Favoured and Mountainous Areas
Mollwaldplatz 5/1, 1040 Wien
Austria

Environmental concerns of mountain farming: an example from the Austrian Tyrol

European mountain areas contain rich natural resources and assets with regard to biodiversity, habitats, landscapes, soils and water. Agriculture has, over long periods, developed and modified this stock of assets reflecting the particular and traditional land-use management forms. In this way, close links have often been developed between environmental assets and specific cultural and lifestyle patterns. In fact, much of the environmental capital can be considered as a contribution to cultural capital, valued through its links to past and current distinctive human activities. However, the systems are quite vulnerable. In mountain areas, one can observe drastic changes in the agricultural structure, such as farm abandonment, afforestation, and transformations in animal husbandry and land-use systems. These might lead to a considerable decline of biodiversity and undesirable landscape changes, in addition to increased natural hazards, forest fires, eutrophication problems, water pollution and soil degradation.

Owing to the specific character and fragile environment of mountain areas, the European Commission launched a project to assess the specific aspects concerning agriculture and the environment in the mountain areas, regarding both constraints and assets as well as the impact of various European Union (EU) and national instruments on environment and mountain farming (Regulations 2078/92 and 2080/92: Objective 1, 5b and 6; market support instruments; community initiatives; etc.). A network comprising six regional European mountain areas and 25 study zones covering the wide range of different physical and socio-economic situations was established under the project leadership of Euromontana in Paris. The Federal Institute for Mountainous and Less-Favoured Areas in Vienna co-ordinated the "Central and Eastern Alps" network.

The "Tyrol Oberland" study area, which is part of this network, is situated in the far western part of the Austrian province of Tyrol. The total size of the area is 335,200 ha, but only 7.1% (23,900 ha) is suitable for permanent settlement. The total population numbers 87,724 (1991), which indicates a low density of just 26 inhabitants per km².

The principal income sources are tourism, energy and the construction industry. Most industrial and trade enterprises are small or medium-sized. They produce mainly for the local market. Tyrol is the most important tourist region in Austria, with one third of all over-night stays and 84% of all jobs in the tourist industry. The study area itself attracts 24.2% of all tourists visiting Tyrol.

About 3.9% of the working population is employed in agriculture and forestry, and almost 83% of all farm managers are part-time farmers. The average size of the agricultural holdings is extremely small and the individual parcels are small and frequently scattered. The majority of farm holdings suffer from difficult natural conditions of high altitude, steep slopes or severe climate. Steep slopes particularly restrict mechanisation and require a great share of manual work. The variety of agricultural products and farming systems is limited. Because of the small size of holdings, the milk quotas and cattle herds are also small, and the output per farm is low in terms of both production and financial turnover. Farm income per agricultural worker is the lowest in Austria, at just 46% of the EU level.

In spite of this difficult situation, there is no significant land abandonment. Most farmers continue to live on a subsistence basis without significant integration into the market economy. Well-established infrastructure facilities (water and energy supply, sewage networks) and excellent road links are decisive in keeping farmers in even the remoter areas. Due to the different forms of natural and economic restrictions, the basic function of mountain farming in this region lies in maintaining and sustaining the landscape and local culture, rather than in profitability.

Though the environmental situation is relatively favourable, particularly when compared with other alpine regions, the balance between environment and agriculture is precarious. It is easily disturbed by difficult climatic and terrestrial conditions or the fragility of socio-economic, demographic and agricultural structures. If farmers withdrew, there would be drastic impacts on:

- the balance of the alpine water system in connection with the safeguarding of fresh water supply and human living space in general;
- the stability of the terrain, particularly on steep slopes;
- biodiversity and landscape;
- equilibrium of natural circulation systems.

In developing strategies to improve the integration of agricultural impacts with the environment, it is essential to stimulate discourse in local society and politics at different levels. The maintenance of landscapes by agricultural activities is to be acknowledged as a prime asset of the cultural heritage. The dynamics of landscape development require integrated concepts which reflect different types of utilisation and interrelations of areas. Structural and regional policy measures must be conceived in close co-operation with tourist development. It is necessary that this process is adapted to the local needs and can rely on a sufficiently flexible set of instruments.

As the alpine grassland in the study area, as in many mountain regions, is still close to environmentally-sound production schemes, the shift to organic farming is an interesting challenge and a bonus for many mountain farmers. In the study area, about 30% of the farmers have turned to organic farming, making use of this specific asset. The continuing ecological trend can be used to introduce regional labels and to launch increased marketing efforts for organic produce.

Direct-marketing and quality production can help to overcome structural problems, but the adaptation of processing and marketing facilities is a prerequisite. Improved marketing of products requires co-operation among farmers. Pluriactivity is of crucial importance in most alpine regions. The objective is to work on an improved interface between farm resources, local culture, industries, tourism and trade. In addition, direct payment schemes are best suited to provide incentives for the survival of small-scale mountain farmers and for sustaining the environment in mountain areas. The support schemes must acknowledge regional peculiarities and the different components of the environment.

Will Williams[1], D.B.A. Thompson[2], M. Yeo[3], P. Corbett[4] and J. Hopkins[5]
[1]English Nature, Northminster House, Peterborough PE1 1UA, UK
[2]Scottish Natural Heritage, 2 Anderson Place, Edinburgh EH6 5NP, UK
[3]Countryside Council for Wales, Plas Penrhos
Ffordd Penrhos, Bangor, Gwynedd LL57 2LQ, UK
[4]Environment and Heritage Service, Commonwealth House
35 Castle Street, Belfast BT1 1GU, UK
[5]Joint Nature Conservation Committee, Monkstone House, Peterborough PE1 1UA, UK

Mountain areas in the United Kingdom: nature conservation issues and opportunities

Uplands and mountains cover around 30% of the UK land surface (Usher and Thompson, 1988). Most of this is managed for agriculture, forestry and game management, and widely used for tourism and recreation. The landscape is predominantly semi-natural; only a few of the more mountainous parts and some of the peatland landscapes are regarded as near-natural.

Past management has had a major impact on the habitat resource and its wildlife but the potential remains for achieving a high degree of sustainable management (e.g., English Nature 1997a; Thompson *et al.*, 1995). Many environmental measures are being adopted by the UK government's conservation organisations and others to protect and enhance the nature conservation interests of the UK uplands. These are largely channelled through *Biodiversity. The UK Action Plan* (Anon, 1994), implementation of the European Commission's (EC) Habitats Directive (Brown *et al.*, 1998), and wider countryside measures largely influenced through agricultural departments (e.g., English Nature 1997a, b; Thompson and Horsfield, 1997). New measures include proposals to introduce National Parks to Scotland, further development of the existing English Nature Wildlife Enhancement Scheme in England, an extended agri-environment scheme in Wales, and proposals for the integration of rural development and agri-environment schemes.

The UK Conservation Agencies currently work to six priorities for the conservation of the UK uplands:

1. achieving positive and sustainable management for nature, especially for the network of protected areas, focused on the favourable conservation status of habitats and species under the EC Habitats Directive, and on Biodiversity Action Plan targets;
2. influencing Government policies, in particular those relating to agriculture, especially to reduce grazing pressures to benefit biodiversity;
3. promoting best practice for habitats and species management;
4. integrating the farm business with nature conservation requirements;
5. increasing public understanding and awareness of required action, and promoting the appreciation and enjoyment of the upland countryside; and

6.   research and monitoring guided by *Biodiversity. The UK Action Plan* and the EC Habitats Directive.

REFERENCES

Anon (1994) *Biodiversity. The UK Action Plan.* The Stationery Office, London, UK.

Brown, A.E., Burn, A.J., Hopkins, J.J. and Way, S.F. (eds.) (1998) The Habitats Directive: Selection of Special Areas of Conservation in the UK. Joint Nature Conservation Committee, Peterborough, UK.

English Nature (1997a) Managing the English Uplands: Putting English Nature's upland strategy into action. English Nature, Peterborough, UK.

English Nature (1997b) Beyond 2000. English Nature's strategy for improving England's wildlife and natural features. English Nature, Peterborough, UK.

Thompson, D.B.A., Hester, A.J. and Usher, M.B. (eds.) (1995) *Heaths and moorland: cultural landscapes.* The Stationery Office, Edinburgh, UK.

Thompson, D.B.A. and Horsfield, D. (1997) Upland Habitat Conservation in Scotland: a review of progress and some proposals for action. *J. Botanical Society of Scotland,* **49**: 501–516.

Usher, M.B. and Thompson, D.B.A. (eds.) (1988) *Ecological Change in the Uplands.* Blackwell Scientific Publications. Oxford, UK.

Jan Kalvoda[1], William Heal[2] and Anthony Dore[3]
[1] Department of Physical Geography and Geoecology, Charles University (Prague)
Albertov 6, 12843, Praque 2, Czech Republic
[2] 1 Whim Square, Lamancha, West Linton, Tweedale EH46 7BD, UK
[3] Meteorological Research Flight, Building Y46, DERA Farnborough
Hampshire GU14 6AP, UK

Introduction: Central and Eastern European mountains

The regional workshop on Central and Eastern European mountains was attended by more than 20 people. Covering a broad spectrum of disciplines, the presented topics included changes in biodiversity, natural hazards and risks, problems deriving from severe air pollution, deforestation and socio-cultural marginality.

The countries of Central and Eastern Europe are currently undergoing a period of rapid societal and economic change. The transfer from socialist state control to the private sector is transforming industry, land-use and management. Changes are occurring in population distribution and landscape structure, causing, among other things, 'architectural' pollution. An important feature of the region is that political boundaries divide many of the mountain ranges. This has major implications for land and water management and for the establishment of national parks and communications. A major effort is required to improve trans-boundary cooperation and to develop interdisciplinary approaches to research in these mountain areas.

The main priority is to continue research on the nature of the mountain ranges, especially as a methodological background for registering and observing current regional and global changes. This is necessary from the scientific point of view, as well for understanding variations of natural and societal change.

A wide range of factors cause change in the environments of the Central and Eastern European mountains:

1. long-term climatic, hydrological and landform evolution;
2. pressures for land-use change in agriculture and forestry;
3. industries, mining and transport development;
4. increases in tourism and conservation, bringing some demographic changes.

It is this wide range of factors, combined with political change and multi-national involvement, which makes these mountains a unique challenge. Consequences of these individual or interacting drivers of natural and societal changes which have been identified include:

1. increased geodynamic hazards and risks (e.g., slope movements, avalanches and floods);

2. global and regional atmospheric pollution, modification of hydrology including water quality, soil degradation and erosion;
3. composition and genetic shift in vegetation;
4. faunal change, including increased pest outbreaks;
5. composition of human populations, conflict of different interest groups and loss of traditional knowledge.

A wealth of data and research experience related to nature, environment and society in the mountains of the region already exist, some covering very long periods. This provides an unusual opportunity for mountain research. In future, this work should be channelled into correlation and general synthesis of regional data about nature and society in the mountains. Priority should also be given to research and other activities in regarding correlation of Central European mountains north of the Alps (e.g., Sumava and Giant Mountains), and the Carpathian system (e.g., Tatras, Fagarsh Mountains) with the nature, environment and societies of the far eastern European and Asian mountain regions, such as the Ural and Caucasus Mountains.

There is a need for special efforts to promote cooperation between scientists from Central and Eastern Europe and those from Western Europe. In particular, two-way interaction through participation in regional conferences needs stimulation.

Anthony Dore[1], Mieczyslaw Sobik and Krzysztof Migala
Department of Meteorology and Climatology, Institute of Geography
University of Wroclaw, pl. Uniwersytecki 1, 50–137 Wroclaw, Poland
[1]correspondence: Meteorological Research Flight, Building Y46, D.E.R.A. Farnborough,
Hampshire GU14 6TD, United Kingdom

The role of orographic cap clouds in pollutant deposition in the Western Sudety Mountains

Introduction

The occurrence of cap clouds is a commonly observed phenomenon at mid-latitudes in mountainous areas with a maritime climate. The direct deposition of cloud droplets on vegetation can make an important contribution to hydrological and chemical inputs, particularly in forested areas where the vegetation efficiently cleans out the hill cap cloud. Although such clouds do not rain, they can be washed out by rain from upper level clouds. This mechanism is known as the 'seeder-feeder effect'. During winter months, the effect may be even more pronounced as snow crystals, with their large surface area, wash out cloud droplets more efficiently than rain.

The Sudety mountains form the border between Poland and the Czech Republic. The western Sudety mountains can be divided into the Izerskie mountains in the most westerly part and Karkonosze (Giant Mountains) further east. The Izerskie mountains are the first orographic barrier during typical westerly circulation and are exposed to highly polluted air from intensive heavy industry and the combustion of sulphur-rich brown coal in the region of the Polish, German and Czech borders. This has caused severe forest destruction, particularly in the altitude range 800 to 1200 m. asl. A survey of forest health in the Czech Giant Mountains revealed a very uneven distribution of spruce defoliation (Moravcik and Cerny, 1995). Defoliation was greatest in the western part of the mountains and at high-elevation sites exposed to wind.

The remaining forests in this area show strong variations in tree health over very short distances, with distinct forest stand edges that delimit a slowly advancing front of tree destruction. Forest damage in this area may in addition be related to a number of different factors, such as frequency of frosts during the vegetation period and the local history of forestry practice. Hill cap clouds occur in this area on approximately 240 days a year (Sobik and Migala, 1993). In a comparable region in North America, Weathers et al. (1995) noted significantly greater cloud deposition to trees at the edge of a spruce stand in the Catskill Mountains. The deposition of ions was on average three times greater at the edge than that on trees in the interior.

Cloud water deposition experiment

An experiment was conducted to compare the cloud water input and ion deposition to a forest stand at two sites, one in the Izerskie mountains and one in Karkonosze. The first site was a forest stand on the northwest slope of Smrek in the western Izerskie mountains at an altitude of 1040 m. Eleven buckets were placed below the forest canopy to measure the amount of throughfall (water, precipitation or cloud water, dripping off the forest canopy to ground) at trees on the northwest edge of the stand and at different distances of up to 60 m from the edge. A similar design was used at an altitude of 1060 m on the northwest slope of Mount Szrenica in Karkonosze, 20 km southeast of Smrek. Measurements were made on a daily or twice daily basis from 19 May to 2 June 1997. Fourteen individual precipitation and cloud deposition events occurred during the project. Despite considerable scatter in the data, a clear trend emerges with the greatest throughfall at Smrek (400 mm) occurring at the collectors near the edge of the forest stand, and decreasing by a factor of about 3 or 4 at the collectors furthest from the forest edge. Much lower values of throughfall (100 mm) were recorded at Mount Szrenica in Karkonosze. These observations indicate a much higher level of cloud base at Szrenica than at Smrek.

Snowfall experiment

During the snow-chemistry project, samples of daily precipitation events were collected between 27th January and 12th February 1994 at four sites at different altitudes. The collectors have a circular dish with a thermostatically-controlled heating device. Precipitation melts on impact and is collected in a bottle. Rime samples (frozen cloud water) were collected on nylon wires. Concentrations of ions in rime were consistently higher than those in snow by factors of 1.3, 1.5, 2.2, 2.8, 2.0, 4.1, 2.6, 2.3 and 1.8 for H^+, NH_4^+, Cl^-, Na^+, K^+, Ca^{++}, Mg^{++}, Fe^{+++}, and Pb^{++} respectively. Measurements of the total depth of snow in different areas were made using probes. The results showed that, for sites of equivalent altitude, the snow depth tended to be greater in Karkonosze than in the Izerskie mountains.

Rainfall experiment

Rain-chemistry projects took place during the spring of 1994 and 1995. Seven rain collectors (consisting of a simple funnel and bottle arrangement) were installed at different sites in the mountains. Cloud water was collected on a daily basis using passive collectors consisting of an array of strung nylon wires.

Total precipitation for the monitoring period May – October 1995 was greatest at the high-altitude sites. Pollutant deposition also increased with altitude, and was greatest in the westerly reaches of the mountain range. Cloud-water concentrations were significantly higher than those in rain, on average by factors of 1.8, 3.1, 3.7, 1.4 and 2.9 for acid, nitrate, ammonium, chloride and sulphate respectively. All ion species had lower concentrations in cloud water at Karkonosze than in the Izerskie mountains.

Numerical modelling experiments

Calculations of orographic precipitation were performed using numerical models which simulate the formation of orographic cap cloud due to the condensation of rising air and cloud-droplet scavenging by raindrops (Carruthers and Choularton, 1983) and snow crystals (Choularton and Perry, 1986).

The modelled rainfall rate was found to reach a maximum over the summit of Smrek with a secondary, smaller rainfall peak occuring over Szrenica. At this point, the heavy precipitation significantly depleted the liquid water in the cloud, so that the seeder-feeder effect was less prominent. The highest chemical deposition rates occurred on the upwind side of the first hill peak and by the time the air reached Szrenica, most of the chemicals had been washed out of the cloud. The first peak in a chain of hills therefore tends to take the highest pollutant deposition from rainfall, and thus effectively 'protects' the successive peaks.

The low terminal velocity of snow crystals means that snow is transported over large horizontal distances. The precipitation peak over Smrek for the orographic snow simulation was therefore broader than in the case of rain, and reached a maximum several kilometres downwind of the hill summit. For snow, transport of pollution occurs over much longer distances than in the case of rain, with deposition of chemicals at Szrenica being almost as high as at Smrek.

REFERENCES

Carruthers, D.J. and Choularton, T.W. (1983). A model of the seeder-feeder mechanism of orographic rain including stratification and wind drift effects. *Q. J. Roy. Met. Soc.*, **109**, 575–588.

Choularton, T.W. and Perry, S.J. (1986). A model of the orographic enhancement of snowfall by the seeder-feeder mechanism. *Q. J. Roy. Met. Soc.*, **112**, 335–345.

Moravcik, P. and Cerny, M. (1995). Forest die-back affected regions of the Czech Republic. In: Acidification in the Black Triangle Region, 11–18, 5th International Conference on Acidic Deposition Science and Policy, Göteborg, Sweden, 26–30 June 1995.

Sobik, M. and Migala, K. (1993). The role of cloudwater and fog deposits on the water budget in the Karkonosze mountains. *ALPEX Regional Bulletin*, **21**, 13–15.

Weathers, K.C., Lovett, G.M. and Likens, G.E. (1995). Cloud deposition to a spruce forest edge. *Atmos. Envir*, **29** (6), 665–672.

Irina Glazyrina
Institute of Natural Resources, Siberian Branch of Russian Academy of Sciences
16 Nedorezova str., Chita 672 014
Russia

The ecological debt concept: implementation for biodiversity conservation in Russia

The ecosystem approach in natural resource management is usually considered as a necessary process towards sustainable development. However, there are several versions of this principle. In this paper, ecosystem approach is defined as the requirement of responsibility for ecosystem integrity, which must be expressed in decision-making procedures in natural resource uses. This responsibility should be recognised for ecosystems of all levels and supported by proper natural resource management and land-use regulation.

If we discuss biodiversity conservation within frameworks of ecosystem integrity, we are faced with the problem of comparative measurement of non-uniform factors of influence and problem of incommensurability. The ecological debt concept gives an opportunity to overcome these difficulties. Ecological debt has been defined by Gofman and Ryumina (1994) as follows:

> "The statement of nature is called initial if it has a capability to full self-restoration. A difference between current and initial states is equal to ecological debt from society to nature".

Ecological debt might be defined in physical and/or monetary terms. Both these expressions have their merits and demerits. However, ecological debt is a dynamic quantity. It depends on human values, which are permanently changing; current scientific knowledge; cultural traditions; and existing technologies. Therefore, the calculation of ecological debt is interdisciplinary in essence. It will be a result of the achievement of an "equilibrium" of different human values, rather than a purely scientific result (Soderbaum, 1994).

In many situations, some part of ecological debt might be calculated exactly at a fixed moment in time. For example, if we consider a forest harvesting project, we can estimate the cost of forest restoration with existing technologies. Other parts might be less well defined. For instance, the restoration of the environment with certain species may require additional research. Finally, the problem remains of whether we should restore initial ecosystems or create an alternative valuable and desirable environment. There is not always a unique solution. Cultural and religious aspects, which are not computable, might be influential. Different generations might choose different ways.

We should take this uncertainty into account. At the same time, imperfection of knowledge is not a reason to ignore the existence of ecological debt. The first step from recognition of ecological debt to its compensation might be established as an 'Ecological Debt Line' within a regional State Yearly Budget: a procedure that is theoretically similar to including environmental costs within National Accounts. A second step might be the creation of financial flows to cover ecological debt. Under current economic conditions in Russia, where the state is still the owner of almost all important natural resources, it is reasonable to combine the "polluter pays" and "victim pays" principles (Golub and Strukova, 1995) in these mechanisms.

The exploitation of natural resources is the only source of economic growth in parts of Eastern Siberia (Ivanov *et al.*, 1996). It currently seems almost impossible to organise "full reimbursement" of ecological debt for these areas. However, recognition of the ecological debt of society toward nature, and its "legitimisation" is now more important for a long-term strategy of economic and social development than its total return or exact accounting. The creation and improvement of procedures to provide a static ecological debt – or at least as close to this as possible (as well as its accounting) – is an important interdisciplinary field of research.

Procedures of ecological debt compensation need the establishment of new institutional instruments and tools. The increase of state influence on economic activities will probably follow. It is important that all these changes should not destroy efficient mechanisms of decentralised markets. Some studies with mathematical models show that constrains for resource stock available for use might be compatible with economic efficiency (Glazyrina, 1997). Minimisation of ecological debt, as an objective function, might be the basis for decision-making procedures at micro-economic level.

Financial support for this study by the Eurasia Foundation and the Russian Humanitarian Scientific Foundation is gratefully acknowledged, as also is support to the author by the John D. and Katherine T. MacArthur Foundation.

REFERENCES

Glazyrina, I. (1997) Edgeworth's conjecture in atomless economies with a non-separable commodity space. *Journal of Mathematical Economics* **27** (1): 79–90.

Golub, A.A. and Strukova, E.B. (1995). *Economics of Nature Use.* Aspekt-Press, Moscow. (In Russian)

Gofman, K.G. and Ryumina, E.V. (1994). Credit relationship between society and nature. *Economics and Mathematical Methods* **30**: 17–30.

Ivanov, B., Glazyrina, I.P., Strizhova, T.A., Vozmilov, A.M. and Zadorozhny, V.F. (1996). Characteristics of the sustainable-development model for the outlying and contact zones of the Lake Baikal basin. In: Sustainable Development of the Lake Baikal Basin. Koptyug, V. and Uppenbrink, M. (eds.) *NATO ASI Series*, 2. *Environment*, **6**: 127–132

Soderbaum, P. (1994). Actors, ideology, markets. Neo-classical and institutional perspectives on environmental policy. *Ecological Economics* **10** (1): 47–60.

M. Herman and Frantisek Zemek
Institute of Landscape Ecology, Academy of Sciences of the Czech Republic
Na sadkach 7, 370 05 Ceske Budejovice
Czech Republic

Integrating the natural and socio-economic potentials of regional landscapes within GIS: an extended base for strategic analysis and planning in the Šumava region of the Czech Republic

Aims

In this study of the Šumava region of the Czech Republic, we focus on two objectives. First, to explore what kind of geographical information systems (GIS) and remote sensing technology information can be used to understand the historical development of a mountain region and its carrying capacity. Second, to investigate ways in which socio-economic and "soft" data can be integrated into a GIS database frame and modelling, to develop useful information for strategic analysis, land cover changes and recent socio-economic potentials of the study region.

Introduction

The first objective has been well addressed by many authors, since digital image analysis systems for processing remotely sensed data and GIS are recognised as basic tools for gathering geographical or spatial information about landscape phenomena at the regional scale. They are well documented both independently (e.g., Richards, 1986; Burrough, 1991) and integrated together (e.g., David, 1991), including state-of-the-art reviews of their utilisation in landscape ecology (Turner et al., 1990; Haines-Young et al., 1993). The second issue, especially the potential of GIS as a process for modelling and simulation modelling, has only recently been addressed (e.g., Eastman et al. 1995, 1997).

Study area

The Šumava Region, a 2,600 km² area in the southwestern part of the Czech Republic, was our study area. It includes the Šumava Mountains, a massive mountain range with an average altitude of 1150 m. and part of the Šumava piedmont. This area has experienced several tides of human migration and is economically marginal at present.

Data and methods

Two digital "forest masks" of the study area were overlaid to estimate long-term land-use change (forest and non-forest classes) between 1830 and 1990, and two satellite Landsat 5 Thematic Mapper scenes (1987, 1995) were processed using a neural net classifier to evaluate both recent land-use (eight classes) and changes after 1989, when a significantly new political situation arose. The calculation of recent socio-economic potential was based on spatial multi-criteria evaluation (MCE) modelling, village cadaster data from the 1995 Czech Statistical Yearbook and a questionnaire survey of local key stakeholders and experts. We used 64 parameters, grouped into seven data categories: demographic characteristics; transport facilities; technical facilities; public services; private and state organisations; tourism facilities; and cultural facilities.

All GIS and imagery data were processed on PC ARC/INFO, v.3.5; PC PCI, v.6.0.1; and IDRISI Win v.2 systems.

Results

The results reveal 20% afforestation over the past 160 years. The current population density is equal to that of 1960, but 65% of that of the 1830s in the Šumava piedmont, and 45% in the Šumava Mountains. The main reasons for this are:

- a decline of small glass factories based on fuel-wood in the nineteenth century;
- establishment of extensive military training areas and emigration of the ethnic German population after 1945;
- founding of state co-operative farms after 1950, which utilised only the land capable of being farmed with 1950–60s vintage Czech farm machinery.

The assessment of land-use changes between 1987 and 1995 shows a 7% increase in meadows and pastures on account of new arable land. This is connected with the privatisation process after 1989 and the abolition of one military camp, causing an abandonment of about 2% of the area, with its subsequent natural regeneration. A total of 2.7% of forest cover was lost, in a clearcut area where spruce was harvested.

The "partial" model outputs demonstrate unevenly distributed socio-economic conditions within categories. But, assuming a full trade-off between categories, and including stake holders' opinion, as weights in the Ordered Weighted Average procedure, it can be concluded that, even if there are differences in socio-economic potential between cadasters, there is no significant difference between mountain and piedmont areas. Calculated on a per person basis, the potential is higher in the Šumava Mountains. There was no relationship found between socio-economic potential and recent or long-term land-use changes.

The above results, together with other GIS layers and ancillary data, have become the foundation for the formulation of alternative scenarios for the future development of the Šumava region.

REFERENCES

Burrough, P.A. (1991) Principles of geographical information systems for land resources assessment. *Monographs on Soil and Resources Survey*, No. 12., Oxford.

David, F.D. (1991) Environmental analysis using integrated GIS and remotely sensed data: Some research needs and priorities. *Photogrammetric Engineering and Remote Sensing,* **57** (6): 689–697.

Eastman, J.R., Jin, W. *et al.* (1995) Raster procedures for multi-criteria/multi-objective decisions. *Photogrammetric Engineering and Remote Sensing.* **61** (5): 539–547.

Eastman, J.R. (1997) *Idrisi for Windows Manuals.* Clark University, Worcester. USA.

Haines-Young, R., Green, D.R. *et al.* (eds.) (1993) *Landscape Ecology and GIS.* Taylor & Francis, London.

Richards, J.A. (1986) *Remote Sensing Digital Image Analysis.* Springer-Verlag, Berlin.

Turner, M.G., Gardner, R.H. *et al.* (1990) *Quantitative Methods in Landscape Ecology – The Analysis and Interpretation of Landscape Heterogeneity.* Springer-Verlag, Berlin.

Marián Janiga
Tatra National Park Research Centre
059 60 Tatranská Lomnica
Slovak Republic

The lead cycle in the Alpine environment of the Tatra mountains: vertebrates as bioindicators

Introduction

Long-range transport still dominates the atmospheric deposition of metals in the cases of cadmium, lead, and mercury (UNECE, 1995). The most important regional environmental problems due to long-range atmospheric transport have been documented regarding the impacts of lead on two types of habitats: forests, especially forest soils, and lakes (UNECE, 1995). Only recent studies have concluded that the lead deposition is positively correlated to precipitation amounts, and consequently to orography. Such an orographic dependence is well documented for air mass barriers like the Alps (BUWAL, 1993; Zechmeister, 1995) or the Tatra Mountains (Šoltés et al., 1992). In this study, mammals and birds were used to monitor lead accumulation in the alpine environment.

Material and methods

The bodies of dead birds were collected from transects along altitudinal gradients on slopes mainly within the southeastern High Tatra Mountains, Slovakia. This study also includes samples of nasal bones from 81 cranial skeletons of chamois. For details on study area, field methods, analytic chemistry, and statistics see Janiga et al. (1997a, b).

Results

Levels of lead in the same cluster of insectivorous-granivorous birds clearly varied with altitude. Lead concentrations in bones of alpine birds (Alpine Accentors, *Accentor collaris*) were higher than in the birds from forest, agricultural, or rural habitats. Lead concentrations were low in bones of birds which are predominantly carnivorous, and high in granivorous or herbivorous birds.

Chamois bone lead levels were strongly increased in the West Tatras compared to the eastern Belianske Tatras. There was also a significant difference between the eastern part of the High Tatras and other regions, including the Low Tatras. In the West Tatras, the levels of lead in the bones of chamois have remained high since World War II; the lowest values tended to be for the period from 1905 to 1946. In general, the concentration of lead in the bones of chamois decreased from west to east.

Discussion

In the Tatras, there is a strong correlation between orography, the amount of wet deposition (Konèek *et al.*, 1973) and rainfall composition (Gazda and Hanzel, 1978). Precipitation increases more or less constantly with altitude and from east to west, being 1.7 times as high in the West Tatras as in the eastern Belianske Tatras (Petroviè and Šteffek, 1967). In high-altitude habitats, westerly and northwesterly winds mainly prevail in the West and Central Tatras (Konèek *et al.*, 1973). Moreover, windspeed may play an important role in the pollution of high-altitude habitats. Heavy metals are usually bounded to submicron particles of aerosols. According to Clough (1975), small particles show a deposition closely related to windspeed. In the Tatras, windspeed increases above the treeline, mainly in the high-altitude alpine areas (Konèek *et al.*, 1973).

Deposition of lead probably initially depends on the element contents of air masses, and it is intensified by high windspeed in the alpine areas. The ratio of elements in the dust found in the Tatras corresponds to the ratio of elements found in some species of mosses (Šoltés *et al.*, 1992), and in general the amount of atmospheric lead tends to be larger than the amount of lead in precipitation water. As the concentration of lead is evidently lower in the precipitation than in gutter water (Tuinský and Chudíková 1991; Šoltés *et al.*, 1992), we assume that particles of lead are brought by air masses and then washed from the rocks to their base. Hajdúk (1988) found that lead content in the soil samples was higher close to rocky walls than in samples from 2 to 8 metres away. The author has also found increased elevations of soil lead in the northern valleys of the West Tatras.

Bednárová & Bednár (1978) found that approximately 50% of lead is physiologically absorbed by the tissues of plants, while the remaining 50% may be washed away. The high positive correlations between lead levels in herbivorous or granivorous vertebrates and lead deposition data are usually explained by the intake of lead from the surface of food plants and the intake of contaminated soil (Froslie *et al.*, 1985). Thus, the geographic pattern of lead concentrations in the bones of Tatran chamois reflects the climatic conditions in the Tatras.

REFERENCES

Bednárová, J. and Bednár, V. (1978) Lead content in the plants of the Tatra National Park. *Zborník TANAP*, **20**: 163–175.
Bundesamt für Umweltschutz, Wald und Landschaft (BUWAL) (1993) Bestimmung der Deposition von Luftschadstoffen in der Schweiz mit Moosanalysen. *Schriftenreihe Umwelt* 194, Bern.
Clough, W.S. (1975) The deposition of particles on moss and grass surfaces. *Atmospheric Environment*, **9**: 1113–1119.
Froslie, A., Norheim, G., Rambaek, J.P. and Steinnes, E. (1985) Heavy metals in lamb liver: Contribution from atmospheric fallout. *Bulletin of Environmental Contamination and Toxicology*, **35**: 175–82.
Gazda, S. and Hanzel, V. (1978) Problems of conserving subterranean waters of the Tatra National Park from the aspect of the contemporary hydrogeological and geochemical knowledge. *Zborník TANAP*, **20**: 183–206.

Hajdúk, J. (1988) Contents of Pb, Cd, As, Fe, Cr, Zn, Cu, Ca, Mg and S in TANAP soils in relation to the effect of industrial immissions. *Zborník TANAP*, **28**: 251–261.

Janiga, M., Chovancová, B., Emberyová, M. and Farkašovská, I. (1997a) Bone lead concentrations in the chamois . *IBEX – Journal of Mountain Ecology*, in press.

Janiga, M., Emberyová, M., Závadská, M., Veselá, A. and Krištín, A. (1997b) Metal pollution seriously threaten the alpine biota – A case study: Lead in the bone tissues of birds from the Tatra Mountains. Submitted to *Environmental Pollution*.

Konèek, M., Šamaj, F., Smolen, F., Otruba, J., Murínová, G. and Peterka, V. (1973) Climatic conditions in the High Tatra Mountains. *Zborník TANAP*, **15**: 239–324.

Petroviè, Š. and Šteffek, M. (1967) Evaluation of measurements recorded on precipitations totalisators in the Tatra region. *Zborník TANAP*, **10**: 105–110.

Šoltés, R., Šoltésová, A. and Kyselová, Z. (1992) Effect of emissions on non-forest vegetation in the High Tatra and Belianske Tatry. *Zborník TANAP*, **32**: 307–333.

Tuinský, L. & O. Chudíková (1991) Precipitation, gutter and gravitation water chemistry in the TANAP forest ecosystems. *Zborník TANAP*, **31**: 97–107.

United Nations Economic Commission for Europe (UNECE) (1995) *Task Force on Heavy Metals Emissions. State-of-the-Art Report*, 2nd Edition. Prague.

Zechmeister, H.G. (1995) Correlation between altitude and heavy metal deposition in the Alps. *Environmental Pollution*, **89**: 73–80.

Jan Jeník
Charles University, Faculty of Sciences
Benátská 2, 128 01 Praha
Czech Republic

Centres of biodiversity and human interference in the middle mountains of Central Europe

The exploration of habitat differentiation and plant and animal diversity in the Central European middle mountains started four centuries ago. Over the intervening period, numerous data with regard to the mountain environment, flora and fauna of the Bohemian Forest, Ore Mountains and Sudetes have been collected and recurrently examined by Austrian, German, Czech and Polish scholars who share their activities in these ranges divided by national boundaries (Price, 1995). Besides the minerals, wood, medicinal plants and game which have been extracted, the natural uniqueness and educational value of mountain plants and animals have been much-appreciated national assets cultivated by the universities and learned societies of Vienna, Munich, Prague, Leipzig and Warsaw. Available data from Schwenckfeld, in 1600; Jirasek *et al.*, in 1791; Sendtner, in 1860; and Šourek, in 1969 enable the description of the successive changes of the biodiversity which is threatened by enhanced human interference, air pollution and the invasion of alien biota.

In contrast to the large and inaccessible summits of high-mountain ranges, the Central European highlands have permitted the collection of many pioneer data relating to altitudinal gradient. In the Giant Mountains, meteorological observations on Sniezka Peak (1602 m) started in 1824, and continuous records have proved the earlier assumptions of notable warming over the past century (Czerwinski, 1995). Long-term observations of wind action, snow and avalanches could be integrated into a theory explaining the overall pattern of biodiversity (Jeník, 1990). In the Bohemian Forest, Sendtner described the chemistry and physics of soils in remarkable detail as early as 1860, and recently these data enabled comparison with soils affected by acid rain. Shortly after the pioneer research of Louis Agassiz in the Alps, in 1882 Partsch confirmed the existence of the former valley glaciers in these middle mountains, and thus opened new reasoning in biogeography and ecology.

In the second half of this century, geologists, climatologists and ecologists jointly explained the ecosystem organisation of mountain habitats, and described the unique environments of tors, scree, rock faces, corries, glacial lakes, patterned grounds, peatbogs, arctic-alpine tundra, krummholz zone and mountain taiga (Jeník, 1990; Soukupová *et al.*, 1995; Vacek, 1996), which coincide with the prominent centres of biodiversity.

Vascular plants remain a favoured object of scientific study, and the long-term history of botanical findings is both a remarkable testimony to scientific methodology, and a tool for monitoring the environmental changes taking place in this region. Šourek, in 1961, compiled a remarkable flora from the Giant Mountains, and a team of Bavarian and Czech botanists is currently preparing an integrated volume of vascular plants covering the entire area of the bilateral Bohemian Forest. Vertebrate animals were similarly well studied in the past, and present-day monitoring of endangered mammals, birds, reptiles and amphibians now easily crosses the national boundaries which served as the "Iron Curtain" for several decades.

Due to the good representation of botanists and zoologists in the universities and learned societies, the akaryotic organisms, fungi and invertebrate animals, have also been intensively studied on the Czech and Polish sides of the Central European middle mountains. The recent development of molecular biology did not restrain classical taxonomy, which was always closely linked with the field research. Also, the Zürich-Montpellier School of phytosociology contributed to the improved inventory of vascular and cryptogamic plants, and the identification of centres of biodiversity. Specialised botanical and zoological journals, and volumes devoted to the ecology and biodiversity of particular middle mountain ranges endure: for example, the 33 volumes of *Opera Corcontica* from the Giant Mountains, and the newly established series *Silva Gabreta* for the Bohemian Forest. The identification of critical taxa, description of new species, compilation of species lists and distribution maps, and elaboration of red-data lists have been supported by recent biodiversity projects sponsored by the Global Environmental Facility and UNESCO's Man and Biosphere (MAB) Programme.

High numbers of indigenous species and many biogeographically rare plant and animal populations occur in the above-mentioned treeless habitats: in the area of glacial corries such as the Gross Arber-Lake and Cerné Lake corries in the Bohemian Forest; Zechengrund in the Ore Mountains; and Sniezne Kotly, Kotelné Jámy and Velká Kotlina in the Sudetes. The large peatbogs also host great numbers of algae, bryophytes and invertebrates. Available censuses and taxonomical records suggest exceptional species-richness of all life forms in these ecological "islands" surrounded by closed-canopy forests. Their origin derives from (a) latitudinal situation between the Alps and Scandinavia, (b) longitudinal position between oceanic and continental Euroasia, (c) altitudinal marginality with regard to the summit arctic-alpine tundra and subalpine forests, and (d) continued absence of the closed-canopy forest during the postglacial succession. Migration routes between biogeographical zones, establishment and preservation of outlying populations (relics), and micro-evolutionary processes resulted in this mixture of contrasting plant and animal populations.

Direct economic exploitation of rare plants (medicinal and ornamental plants) and animals (specimens highly prized by collectors) in the centres of diversity is no longer a problem, as most are protected as nature reserves. However, major damage is caused by human-induced changes of physical environment, spontaneous invasion of alien species, and overpopulation of red deer. Besides the air pollution in the infamous "Black Triangle", the middle mountains of Central Europe suffer from the increasing demands of visitors seeking recreation, tourism and sports. Some of the established nature reserves, national parks and biosphere reserves face an increasing political impact in demands for new sports venues, and even for free access to highly-valued centres of biodiversity. However, due to the new political situation in Central Europe, transboundary co-operation in conservation has steadily improved.

REFERENCES

Czerwinski, J. *et al.* (1995) Wysokogórskie obserwatorium meteorologiczne na Sniezce. Instytut Meteorol. i Gospodarki Wodnej, Wroclaw.

Jeník, J. (1990) Large scale pattern of biodiversity in Hercynian massifs. In Krahulec, F. *et al.* (eds.), Spatial processes in plant communities. *Academia*, Praha, 251–259.

Price, M.F. (1995) *Mountain Research in Europe*. UNESCO/Parthenon, Paris/Carnforth.

Soukupová L. *et al.* (eds.) (1996) Arctic-alpine tundra in the Krkonoše, the Sudetes. *Opera Corcontica*, **32**, 5–88.

Vacek S. (ed.) (1996) Monitoring, research and management of ecosystems in the Krkonoše National Park region. Výzk. ústav les. hosp. a myslivosti, Opoèno.

Tatyana E. Khromova and Ludmila P. Chernova
Institute of Geography, Russian Academy of Sciences
Staromonetny 29, 109017 Moscow
Russia

Using GIS mapping to estimate snow and glacier change in mountain regions (the Alps and the Caucasus)

Mountains and high plateaux occupy approximately 20 percent of the global land surface, and are the source of significant water resources. Some 25 million people reside permanently at elevations of 3,000 m or above, and millions more visit high elevations for summer and winter tourism and sports. Hence, knowledge of the present climate regimes of mountain regions, and an understanding of their variability and sensitivity to change, are of vital importance to scientists in many disciplines. In view of their enormous economic importance for winter tourism as well as for water supply, records of snow cover and glaciers in mountain areas have begun to receive considerable attention.

The development of geo-computer science provides new opportunities for the management of huge volumes of the information from all areas of geographical knowledge, including the science of snow and ice in plains and mountains. These opportunities have become a reality through the central role of the cartographic model, the basic tool of geographical researchers, in the structure of geographic information systems (GIS). Twenty years ago, the creation of the *World Atlas of Snow and Ice Resources* began, based on recent geographical information and current theories of glaciology. The main purpose of the Atlas is to provide global and regional estimates of snow, ice and meltwater resources, and to characterise regime, variability and prospects for the utilisation of natural ice (Kotlyakov *et al.*, 1997).

Natural ice is of great importance for climate, fluctuations of sea level, and river runoff. It is critical for power production, irrigation of arid areas, transportation of great amounts of water, construction and recreation in polar and alpine regions. For insufficiently studied areas, special methods and techniques were developed through comprehensive research, conducted specifically for this purpose. It consequently became possible to compile, for the first time, maps of solid precipitation, temperature conditions, snow cover, and runoff in alpine areas. The maps were compiled with due regard to the indicative role of glacio-nival phenomena.

Over the same period, the wide adoption of system mapping methods and statistical methods, and then electronic mapping, took place in geographical science. Now, automated mapping using GIS has a primary place in geographical studies. Geo-computer science is the synthesis of geography, computer science, theory of information systems and cartography. As a result, we can see and apply the continuous process of map creation and map use in which cartographic models, their transformations, analysis and synthesis form continuous chains.

Glaciology, the part of geography concerning the science of natural ice, was one of those best suited for the application of modern geo-information methods. In the *World Atlas of Snow and Ice Resources*, there are maps of all mountain regions, showing the spatial distribution of winter snow storage and mean summer air temperature. The scales of these maps are from 1:3,000,000 to 1:5,000,000. The maps allow comparison of temperature and moisture in the Alps and Caucasus. It was found that average values are very close. The mean depth of snow storage in the Alps is 1610 mm and 1920 mm in the Caucasus. Mean summer air temperatures are equal at 3,000 m in the two ranges. Thus there are similarities in the climatic condition of glaciers in these ranges.

This allows us to use glacier dynamics as a comparative indicator of climate change in the Alps and the Caucasus. GIS mapping was used to analyse snow and glacier spatial and temporal variability in these mountain areas, principally using data of glacier fluctuation and records of snow cover from meteorological stations. This information was used to construct isoline maps: maps of the annual shift of glacier fronts. Annual front shift data (August to August) were evaluated for 72 Alpine and 13 Caucasus glaciers for the period 1986 to 1990. The resulting maps show generalised spatial data, and indicate that glacier front shift fluctuation varies from year to year and does not depend on glacier size. Comparing the average mapped glacier front shift data with summer air temperature data (ablation condition) and winter precipitation data (accumulation condition), it is concluded that the glacier front shift value reflects the condition of glacier accumulation. Thus, the glacier front shift maps show whether the preceding winter was snowy or not. The maps show that snowy and minor snowy winters occurred in the same years in both the Caucasus and the Alps, offering a means of assessing how snow conditions change in mountain areas.

REFERENCE

Kotlyakov, V. (editor-in chief) (1997) *The World Atlas of Snow and Ice Resources.* Institute of Geography, RAS. Moscow, 1997. 392 pp.

Piotr Migoñ, Krzysztof Parzóch and Roman Zurawek
Department of Geography, University of Wroclaw
pl. Uniwersytecki 1, 50–137 Wroclaw
Poland

Natural hazards in the Sudetes, Central Europe: the human context

Natural geomorphological hazards, such as catastrophic mass movements or floods, are usually perceived as phenomena typical of high mountain environments. Reports of their occurrence in middle mountains are rare, both because of a much lower frequency of events and because their effects are less disastrous, in comparison to the damage they cause in high mountains (Gerrard, 1990). It seems that there is a general tendency to dismiss their importance and interpret them as unpredictable anomalies in an otherwise low-energy environment. Nevertheless, natural hazards are an issue of concern, and it is argued here that the resulting threats become more serious as human interference increasingly transforms physical systems of mass and energy transfer. In this paper, we present some of the natural hazards in the Sudetes of Central Europe (Poland, Czech Republic), and consider various human activities in the mountains and how they contribute to increasing risk.

The Sudetes are a medium-height (600–1600 m), presently unglaciated mountain range, whose morphology is dominated by elevated plateaux and smoothed watershed ridges bounded by steep slopes and dissected by deep valleys. The highest parts of the Sudetes, above 1200 m, are located above the timberline. The Sudetes have a long history of human occupation and related environmental changes. Impacts have been exerted by mining, agriculture and grazing, forestry, industrial and transport infrastructure development, and, most recently, by increasing tourism (Janczak, 1985). Large parts of the Sudetes have recently suffered from catastrophic forest decline, the joint effect of improper forest management practices in the past and high levels of air pollution (Mazurski, 1986). Three categories of natural hazards may be identified as being of concern to the local population: mass movements, floods and avalanches. The former include debris flows and sporadic landslides and rockfalls.

Debris flows are restricted to the highest parts of the Sudetes, where they repeatedly occur on steep slopes of glacial cirques, in amphitheatral valley heads and within deeply incised valleys following heavy rainfalls (Pilous, 1977). Historical records suggest an increasing frequency of debris flows in the 18th and 19th centuries; probably a coupled effect of climate deterioration during the Little Ice Age and the deforestation of mountain slopes. In the 19th century, debris flows were particularly hazardous because of the expansion of human settlements high into mountain valleys. In 1897 in Obrí dul, Karkonosze (Giant Mountains), a debris flow overran a group of dwelling houses, killling seven people. At present, the impact of debris flows is limited although they occasionally affect technical tourism infrastructure, including forest tracks and footpaths, leading to their temporary closure. As both debris flows and snow avalanches usually occur in predictable places, risk zones can be easily mapped. Avoidance of these zones is the best way to mitigate the hazard and this is the policy adopted by the National Park authorities on both the Czech and Polish sides of the Giant Mountains.

The most disastrous consequences for human settlements in the Sudetes result from catastrophic floods, whose frequency has been increasing since the 18th century (Trzebinska and Trzebinski, 1954). The floods in 1854, 1897, 1903, 1938, and, most recently, July 1997 were particularly devastating. The increasing magnitude and frequency of floods, with a peak at the turn of the 19th century, may be causally linked with forest clearance and the associated disruption of the hydrological cycle, settlement expansion and the growing pro-portion of agricultural land. Recent forest decline due to environmental pollution has enabled evaluation of the hydrological consequences of vegetation changes in mountain catchments (Dubicki, 1993); these may be used as a model for earlier phases of forest clearance. They include a 10–16% increase in average yearly runoff; the occurrence of the spring runoff maximum 15 to 30 days earlier; a 30% increase in snowmelt runoff maximum; a 40–90% increase in the height of flood waves following heavy rainfalls; and a 30–95% increase in flood discharges.

Heavy economic losses resulting from floods may be, at least in part, attributed to various human activities and related environmental changes. These include extensive catchment urbanisation, massive deforestation and channel-isation of river courses, all of which have reduced catchment natural storage and contributed to higher runoff velocities. Embankment construction and inade-quate maintenance of small-scale flood reduction measures result in insufficient capacity of river channels and induce flooding of valley floors. Furthermore, the impact of floods might have been less serious if large parts of floodplains and alluvial fan settings had not been converted into built-up areas. Finally, the case of July 1997 flood has shown that some of the emergency actions undertaken were either inadequate or suffered from a lack of coordination at regional and international scale.

The case of the Sudetes confirms that, to a large extent, human-environment relationships in a particular location determine whether a natural phenomenon becomes hazardous or not. Debris flows used to pose more threat than they do now. In contrast, the impact of floods is unlikely to decrease as long as natural environmental conditions in mountain catchments are heavily altered.

REFERENCES

Gerrard, J. (1990) *Mountain Environments.* Belhaven Press, London.

Dubicki, A. (1993) Changes in catchment discharge associated with forest dieback in regions of Poland affected by long-range transported air pollutants. *Ecological Engineering*, **3**: 291–298.

Janczak, J. (1985) Czlowiek i przyroda. Przeglad zmian w srodowisku geograficznym Slaska w ostatnim tysiacleciu. DTSK, Wroclaw.

Mazurski, K.R. (1986) The destruction of forests in the Polish Sudetes Mountains by industrial emission. *Forest Ecology and Management*, **17**: 303–315.

Pilous, V. (1977) Strukturní mury v Krkonošich III. Opera Corcontica, **14**: 7–94, Praha.

Trzebinska, M. and Trzebinski, J. (1954) Zagadnienie powodzi na Dolnym Slasku. *Gospodarka Wodna*, **14** (4): 146–150.

Lenka Soukupová
Institute of Botany, Academy of Science of Czech Republic
252 43 Prùhonice
Czech Republic

Czech national parks: recent human-induced changes in montane taiga, mires and alpine tundra

In densely inhabited Central Europe, agriculture, silviculture, tourism and air pollution have seriously interfered with mountain nature. In order to protect the unique mountain ecosystems which are highly rated because of their particular landforms and biota, two national parks have been established in the Czech Republic: Krkonoše in the Giant Mountains (50°40′N, 15°40′E) in 1963, and Šumava in the Bohemian Forest (48°50′N, 13°40′E) in 1991. These national parks display a pronounced altitudinal zonation, including belts of deciduous broad-leaved forest, montane coniferous taiga, coniferous and broad-leaved krummholz, and arcto-alpine tundra.

Treeless arcto-alpine tundra with patterned grounds, cryoplanation terraces, tors and subarctic mires extends over the windswept and snow-rich summits. It encompasses (i) the cryoeolian zone of convex georelief, with frequent freeze-and-thaw cycles, (ii) the vegetated-cryogenic zone over flat plateaux and saddles, and (iii) the niveo-glacigenic zone of the glacial corries and nivation niches (Soukupová et al., 1995). In the Giant Mountains, pristine "islands" of tundra contain numerous plant and animal species of alpine and subarctic affinity, and represent exceptional elements of biodiversity within the primarily forested landscape of central Europe (Jeník, 1997). In the areas of patterned ground, recent soil-forming processes, current regelation and deposition of snow have been studied by field measurements and mathematical models, in connection with the effects of ill-considered dwarf pine plantations. Comparison with earlier fine-scale ecological maps showed considerable changes in the composition of the plant communities described according to the methodology of Zürich-Montpellier School: the *Cetrario-Festucetum supinae* and *Carici fyllae-Nardetum* have suffered by expansion of the competitive grasses which spread due to increased deposition of airborne nitrogen compounds. A dense net of trails frequented by tourists affected the invasion of alien plant species and induced remarkable cases of genetic erosion, such as in *Viola sudetica* (Krahulcová et al., 1996).

Stands of krummholz consisting of clonal woody species – *Pinus mugo, Salix silesiaca, S. lapponum, Ribes petraeum* – cover large flanks and corries at lower altitudes in the Giant Mountains. Though well-adapted to severe environments, even autochthonous stands of dwarf pine are impaired by the acid rain and suffer from both insect and fungal diseases. In the Bohemian Forest, a pine hybrid *Pinus* × *pseudopumilio* grows in the montane patterned mires situated in the concave landforms above 1000 m. This pine encircles open patches in the centres of the bogs, which display a remarkable variety of semi-terrestrial and aquatic communities adapted to the highly diversified microtopography with hummocks, ridges, lawns, flarks, pools and bog-lakes. Spontaneous development of subalpine mires was marked by only a few changes in the past 60 years. However, severe irreversible changes in both hydrological regime and the chemistry of bogs have been caused by the import of limestone ballast to reinforce the tracks crossing the mires. In contrast, large valley raised-bogs behind the natural levées in floodplains of the Bohemian Forest are occupied by open-canopy woodlands of *Pinus rotundata*, which are particularly species-rich in their central openings and along their marginal zone (Soukupová, 1996). Locally, laggs of the valley raised-bogs have been seriously affected by eutrophication caused by enhanced fertilisation of fields in the catchment area.

Extensive areas of mountain spruce taiga between 900 and 1275 m have controlled the plant life and hydrology of pertinent watersheds since the Atlantic period. Recently, an enormous load of acidifying pollutants has resulted in the dramatic die-back of both autochthonous and planted Norway spruce stands. The main airborne pollutants, SO_2 and NO_x, cause progressive soil acidification and affect nutrient cycling, biomass productivity and interspecies relationships. This has resulted in subsequent changes of the surviving communities, causing species invasion and population explosions of insects and fungi. Vegetation serves as a suitable bioindicator of the changes in the functioning of the ecosystems. Five-year monitoring of soil chemistry, growth of herbs and the development of arbuscular fungi (Vosátka *et al.*, 1995) suggests that after the loss of tree fertility, natural regeneration in acidified forests has been further inhibited by expansion of perennial grasses such as *Calamagrostis villosa* and *Avenella flexuosa*. Spruce saplings and young age classes are also clearly missing, due to the overpopulation of red deer. Afforestation of the large clearings encounters a variety of adverse factors such as the retreat of mycorrhizal fungi, and damage by rodents. Changes in the hydrological role of these forests have affected the hydrology of streams and floodplains in the surrounding landscapes.

In recent decades, botanical and ecological research in these middle mountains has indicated a modest but obvious increase of air temperature, suggested by Dubicki and Glowicki (1994) as about 1°C in the course of the 20th century. Long-distance industrial air pollution (acid rain), excessive eutrophication of streams, irreversible exploitation by forestry and tourism all induce new issues which must be tackled in the near future in order to improve management of the national parks.

The author wishes to acknowledge support from the Grant Agency of the Czech Republic, the World Wildlife Fund and Global Environment Facility.

REFERENCES

Dubicki, A. and Glowicki, B. (1994) Tendencje zmian warunków hydroklimatycznych na obszare Sudetów Zachodnich w biezacym stuleciu. *Práce Instytutu Badaw. Les.*, *ser. B*, 21(1): 85–104.

Jeník, J. (1997) Anemo-orographic systems in the Hercynian Mts. and their effects on biodiversity. Acta Univ. Wratisl. No. 1950, *Prace Instyt. Geogr.*, *seria C, Meteorologia i Klimatologia*, 4: 9–21.

Krahulcová, A., Krahulec, F. and Kirschner, J. (1996) Introgressive hybridization between a native and an introduced species: *Viola lutea subsp. sudetica versus V. tricolor. Folia Geobotanica & Phytotaxonomica*, 31 (2): 219–244.

Soukupová, L. (1996) Developmental diversity of peatlands in Bohemian Forest. *Silva Gabreta*, 1: 99–107.

Soukupová, L., Kociánová, M., Jeník, J. and Sekyra, J. (eds.), (1995) Arctic-alpine tundra in the Krkonoše, the Sudetes. *Opera Corcontica*, 32: 5–88.

Vosátka, M., Soukupová, L., Rauch, O. and Škoda, M. (1995) Expansion dynamics of *Calamagrostis villosa* and VA-mycorrhiza in relation to different soil acidification. In Flousek, J. and Roberts, G.C.S. (eds.), *Mountain National Parks and Biosphere Reserves: Monitoring and Management.* Špindlerùv Mlýn, Czech Republic, pp. 47–53.

Jan Tesitel, D. Kusova, M. Bartos and I. Hanouskova
Institute of Landscape Ecology, Academy of Science of the Czech Republic
Na sadkach 7, 370 05 Ceske Budejovice
Czech Republic

Approaches to overcoming regional socio-cultural marginality: experiences from the Czech Republic

Introduction

The changes taking place in the Czech Republic – and in Central Europe in general – can be characterised as the replacement of the egalitarian principle by the differentiation of development of individual areas according to various socio-economic and natural conditions. The differentiation of development of individual areas, however, can lead to the marginalisation of certain areas, probably accompanied by increased depopulation, with all its socio-economic and ecological consequences (Bartos and Tesitel, 1996). In the socio-cultural context, marginality also includes the impossibility for the majority of the population living in a particular area to integrate itself into the main development trends of the whole society. It reflects specific lifestyle and lifetime strategies.

Viewed from the perspective of further development, such areas have a chance to overcome their marginality by utilising both internal and "external" sources. To start the process, the following conditions need to be met:

- recognition of "comparative advantages" of the area itself;
- local social groups being potential implementers of activities utilising these advantages;
- support to these social groups to realize their potential, mainly in the form of external contributions (state subsidies, foreign investment, international projects, etc.).

Model areas

The marginal areas of the Czech Republic are mainly mountainous or hilly lands where nature is well preserved. The recreational exploitation of marginal areas therefore represents a potential chance to overcome their socio-cultural marginality by diverting it into comparative advantage.

Two marginal areas – the Sumava Mountains (on the border with Germany and Austria) and the Broumovsko region (on the border with Poland) – were the model areas for our study. The comparison of the historical causes of marginality, the description of the present situation and opportunities, and the estimation of their future socio-economic development were the main research issues. The internal potential identification was done mainly for Sumava, while Broumovsko was used to compare the impact of external conditions on socio-economic development.

Identification of internal potential (Sumava)

The development of leisure activities and related services has been generally perceived by the local population as the most promising short-term strategy, while development of spa and sanatorium facilities appears to be the predominant activities for the longer-term "desired future". However, the current form of the tourist industry evidently threatens the continuous existence of both the natural and the cultural resources on which it is based (Moss, 1994; Price *et al.*, 1997). Reaching the point of decline will likely be manifested by the area becoming overcrowded by tourists or holiday makers, i.e., by loss of its genius loci and, consequently, its attractiveness.

The area is also and even more so threatened by "external" factors which are inherent to the nature of tourism. Besides predictable long-term scenarios taking "tourist density" into account, and based on, for example, analyses of expected climate conditions and economic trends, there is a danger of volatile fashion waves which are generally short term and hardly predictable. Such a wave brought a great number of western visitors to Sumava immediately after 1989. The boom is now over.

Tourism and leisure activities, despite their potential, should therefore not be considered as the only means of further regional development. Sustaining the diversity of social groups related to different activities seems to be a necessary precondition to ensuring the flexibility of the social system which, consequently, will strengthen its long-term adaptability to future conditions.

External conditions (comparison of Sumava and Broumovsko)

Despite the fact that the two areas can be considered similar in terms of internal potential, there are many differences between them. This introduced the issue of external forces and their role in the process of regional development. A difference in the "welfare gap" was identified as the most likely reason for these differences. 1995 figures indicate that the welfare gap between the Czech Republic and Germany amounts to 2.51, while that between the Czech Republic and Poland is 1.47 (World Bank, 1996).

In the short term, the German "rich neighbour" offers better paid job opportunities for Czech inhabitants of Sumava and can serve as an example to be followed. Incomes earned by Czech people abroad, as well as money spent on Czech territory by German and Austrian visitors, contribute to the development of "domestic capitalists" who can begin territorial development. The opening of the borders after 1989 immediately plunged the Sumava region into the European context. It is even anticipated that this region is likely to be under pressure from "western" developers who have recognized its role of "empty and relatively underdeveloped" space within densely populated and "sold" Europe. The pressure is already present in hidden form, but the main attack is expected to start after the Czech Republic enters the European Union.

The situation in Broumovsko is different. External pressure generated by neighbouring Poland, providing impetus to the socio-economic development of the area, has been almost absent. Economies on both sides of the border seem to have similar structure and power. Under these circumstances, cross-border economic activities profit mostly on volatile advantages – temporary slight differences in prices of individual commodities – which are generally realised by stall-holders. In the long-term perspective, it is likely that neither economy will have surplus capital to be invested on the other side. From this viewpoint, Broumovsko may play a role of an area being in reserve for future development, yet to be identified. Being perceived and considered in this way, however, it is necessary to expect continuing depopulation of the area, or at least a shift in population structure and the intensification of its marginality.

REFERENCES

Bartos, M. and Tesitel, J. (1996) Large scale land abandonment (problems of reinhabitation). In: Steinberger, Y. (ed.) *Preservation of Our World in the Wake of Change*, Vol. VI. ISEEQS Publication, Jerusalem, p. 871–874

Moss, L.A.G. (1994) Beyond tourism: the amenity migrants, In Mannermaa, M., Inayatullah, S. and Slaughter, R (eds), *Chaos in Our Uncommon Futures*. University of Economics, Turku, pp. 121–128.

Price, M.F., Moss, L.A.G. and Williams, P.W. (1997) Tourism and amenity migration, In Messerli, B. and Ives, J.D. (eds) *Mountains of the World – A Global Priority*. Parthenon, New York, pp. 249–280.

World Bank (1996) *World Bank Yearbook 1996*. World Bank, Washington DC.

Petru Urdea
Western University of Timisoara
Pestalozzi Str. 16, 1900 Timisoara
Romania

The map of the geomorphological risk in the Transfagarasan Highway area (Southern Carpathians, Romania)

Geomorphic hazard is defined by Schumm (1988) as any landform change, natural or otherwise, that adversely affects the geomorphic stability of a place. The building of the Transfagaran highway has also represented a modification of the components of the landscape in the central zone of the Fagaras Mountains (Southern Carpathians or Transylvanian Alps) of Romania. The building of this unique highway in the Carpathians brought solutions for some special geotechnical problems deriving directly from the specific alpine relief, lithology and geological structure. Because the geomorphological component has suffered important modification, including the dynamics of current geomorphological processes, with major implications on the development of activities along this road, it is appropriate to consider the geomorphological hazards in this area. Our research comprised a morphometrical study, a morphological and morpho-dynamical study, and production of a geomorphological risk map.

The "Transfagarasan" highway is 90.167 km long, of which about 30 % is in the alpine zone. The highway crosses the main ridge of Fagaras Mountains – the highest mountains of Romania – between Paltina Peak (2,399 m) and Capra Peak (2,417 m), reaching an altitude of 2042 m. The analysed section lies between Piatra Alba (1,090 m) on the north slope and the confluence of Capra river and Piscul Negru river (1,276 m). The highest altitude of this study area is Vantoarea lui Buteanu peak (2,507 m).

As evaluation of geomorphological risk is closely linked to morphodynamic potential, morphometrical and morphographical data are required. Most of the interfluves comprise typical sharp ridges and pyramidal peaks, some extending above 2,200 m. These interfluves are separated by glacial cirques and valleys with a characteristic U shape. The depth and density of the relief fragmentation – calculated according to Partsch-Krebs method (1922) – are, respectively, about 400–880 m and 3–8,5 km/km². Most of the study area has slopes of over 20°. These morphometrical elements sustain, and at the same time explain, the high morphodynamic potential of the mountain zone crossed by the highway.

The Transfagarasan highway was built between March 1970 and September 1979. A total of 3.8 million m³ was moved, including 212,000 m³ of hard rock and 150,000 m³ of terracing. If we consider that this work changed the profile of the slopes, that new escarpments a few metres high appeared, and that 325 ha of forest were cleared, on both sides of the road, it is clear why the morphodynamic potential of the zone has increased. For morphological reasons, vibrations produced by blasting and heavy machinery resulted in geomorphological events even during the construction period. These included: debris avalanches on 14 June 1972 (7,000 m³), 28 April 1972 (100,000 m³), and 15 April 1973 (8,000 m³); a debris flow on 30 June 1974 (8,000 m³); and an avalanche on 13 March 1973 which was 1.5 km long, 150 m wide, 10 m high, and destroyed 5 ha of forest. Because of natural processes – gelifraction, landslides, rolling avalanches – and traffic vibration, masses of debris have accumulated at the base of slopes along the highway, creating traffic hazards. Our observations and measurements show that this debris amounts to 0.5–3.5 m³/linear metre, compared to 0.1–0.7 m³ under natural conditions.

In the study area, as well as elsewhere in the Romanian Carpathians (Balteanu, 1997), avalanches, rockfalls, landslides, debris flows, gullying, and torrential erosion are major risks. Starting with the evaluation of the morphodynamic potential of this area, a geomorphological evaluation, based on morphometrical, morphographical, and morphological analyses, was used to produce a risk map with four risk zones.

The zone without risk – or zero risk – is restricted to the smooth interfluve areas in Paltinu-Picioru Paltinului, in the central part of Capra cirque, and the largely rounded, well-forested interfluve areas.

The zone with a low level of risk is represented by areas with diffuse phenomena, for two reasons: low morphodynamic potential, as in the case of areas which are smooth or have a low slope angle in cirques and glacial valleys, or on the alpine zone interfluves; or a good forest cover, unaffected by human intervention. If such areas are disturbed, this may result in a higher potential risk for critical meteo-climatic situations, e.g., heavy rain- or snowfall.

The zone with a medium risk is characteristic for the lateral part of axial areas of the valleys and glacial cirques, which are affected by strong vertical fluvial erosion, rockfalls, landslides and even by avalanches over glacial rock steps. Close to the bases of slopes, the mobilisation of debris masses through the phenomena of debris flow or gullying processes are the most important elements that disturb the geomorphological balance.

The zone with a high level of risk includes the majority of the study area, characterised by a high morphodynamic potential. The sharp interfluves, with crests above very steep slopes, and continued by cirques and the walls of glacial valleys, are the main locations where the risks of avalanches, the rockfalls, debris flows and vertical erosion are very high.

The Transfagarasan highway is in an area with a high morphodynamic potential and because of this, most of the area is characterised by a high level of geomorphic risk. Building the highway exaggerated the morphodynamics of the area. Future actions such as, for example, forest clearing on the neighbouring slopes, might lead to rapid and serious uncontrollable evolution.

REFERENCES

Balteanu, D. (1997) Romania. In Embleton, C. and Embleton-Hamann, C. (eds), *Geo-morphological Hazards of Europe, Developments in Earth Surface Processes.* Elsevier, Amsterdam, pp. 409–427.

Schumm, S.A. (1988) Geomorphic hazards – problems of prediction. *Zeitschrift für Geomorfologie*, Supplement 67: 17–24.

Frantisek Zemek and M. Herman
Institute of Landscape Ecology, Academy of Science of the Czech Republic
Na sadkach 7, 370 05 Ceske Budejovice
Czech Republic

Forest decline and the environmental factors in the "Black Triangle" of Central Europe

Dramatic forest die-back has occurred in Central European mountains in recent decades, especially in the Norway spruce forests. There have been many studies of this ecological problem, both to develop methods for the detection and monitoring of forest changes (e.g., Lambert, Ardö et al., 1995; UNEP, 1995; Ekstrand, 1996; Ardö, Lambert et al., 1997; Müller-Edzards, Erisman et al., 1997) and to determine reasons for the decline (e.g. Kandler and Innes, 1995; Reuther and Akgöz, 1995).

This study focused on a geographical information system and satellite-based assessment of forest-cover change from the 1970s to the 1990s in one of the most polluted European mountain areas, often called the "Black Triangle", and used these techniques to estimate the relationship between the changes and some natural and anthropogenic stress factors. The study area covers four mountain ranges and their foothills, namely the Slavkovsky les mountains, the Krusne Hory mountains, the Jizerske Hory mountains, and the Krkonose mountains, an area of 12,000 km^2 in the north-west of the Czech Republic.

Two sets of forest-cover maps with five spruce classes of varying health status, pine, mixed and broad-leaf forests, and deforestation (clear cuts of damaged or dead trees) were generated from satellite and ancillary digital elevation data using supervised neural net classifier and field verification. The 1970s set was derived from two scenes of Landsat multi-spectral scanner (MSS) data, and that of the 1990s from three scenes of Landsat 5 thematic mapper (TM) data. Environmental factors are represented by altitude; climatic conditions; soil type; sulphur deposition; concentration of SO_2; and base cation (BC) release rate from the weathering of primary and secondary minerals of bedrock.

The comparison of classification outputs (an 82% overall accuracy) between the 1970s and 1990s showed a significant increase of highly damaged and deforestation areas in three mountain ranges: 16% increase in the Krusne Hory, 20% in the Jizerske Hory, and 22% in the Krkonose. There is also a 21% shift from healthy and slightly damaged classes to a moderately-damaged class. The following conclusions about the influence of certain environmental factors on spruce forest status are based on a comparison between the data on spatial distribution of forest classes and maps of those factors.

Higher altitudes have a negative influence on spruce development, mainly in the Jizerske Hory and Krkonose mountains. The results reflect the different orographical conditions of these mountains. While the most damaged classes and deforestation in the Jizerske Hory mountains are found in the upper plateau, stand conditions deteriorate more with elevation in the Krkonose mountains. These conclusions are also supported by climatic data. The most damaged spruce classes and deforested areas in the Krkonose and Jizerske Hory mountains lie in areas of 'low summer and high winter index', according to Müller-Edzards, Erisman *et al.* (1997).

Stands of Norway spruce on Mollic Cambisol, Eutric Cambisol, Distric Cambisol and Distric Acid Cambisol soils display less damage than those on Cambic Podzol, Ferro-humic Podzol and Distric Histosol soils in all studied areas. A positive correlation was observed between the degree of spruce damage and SO_2 concentration and sulphur dry deposition in the Krusne Hory and Jizerske Hory mountains. The lowest BC release is in the Jizerske Hory mountains, but there are no differences in BC release under different forest classes there. Low BC saturation need not be a primary reason for forest die-back in the Krusne Hory Mountains, as higher BC release was found in more damaged stands. The highest BC release is in the Krkonose mountains: twice as high as in the Jizerske Hory Mountains. This may have contributed to later spruce decay in comparison to the Jizerske Hory mountains.

REFERENCES

Ardö, J., Lambert, N. *et al.* (1997) Satellite-based Estimations of Coniferous Forest Cover Changes: Krusne Hory, Czech Republic 1972–1989. *AMBIO*, **26** (3): 158–166.

Ekstrand, S. (1996) Landsat TM-based forest damage assessment: Correction for topographic effects. *Photogrammetric Engineering and Remote Sensing*, **62** (2): 151–161.

Kandler, O. and Innes, J.L. (1995) Air pollution and forest decline in central Europe. *Environmental Pollution*, **90** (2): 171–180.

Lambert, N. J., Ardö, J. *et al.* (1995) Spectral characterisation and regression-based estimates of forest damage in Norway Spruce stands in the Czech Republic using Landsat Thematic Mapper data. *International Journal of Remote Sensing*, **16** (7): 1261–1287.

Müller-Edzards, C., Erisman, J.W. *et al.* (1997) Ten years of Monitoring Forest Condition in Europe: Studies on Temporal Development, Spatial Distribution and Impacts of Natural and Anthropogenic Stress Factors. Technical Background Report. Brussels, Geneva, EC and UN/ECE.

Reuther, M. and Akgöz, E. (1995) Forest condition assessment in the Fichtelgebirge and Erzgebirge using remote sensing and GIS technology. EARSeL workshop: GIS and pollution monitoring, Brandys nad Labem, Czech Republic, EARSeL.

United Nations Environment Programme (UNEP) (1995) Forest Damage in Central European Mountains – Final Report of a Large Area Operational Experiment for Forest Damage Monitoring in Europe Using Satellite Remote Sensing. UNEP, Geneva.

Bernard Debarbieux
Institut de Géographie Alpine, Université Joseph-Fourier
17 rue Maurice Gignoux, 38031 GRENOBLE cedex
France

Introduction: Southern European and Mediterranean mountains

Although the Mediterranean basin is a very mountainous area, it is not the most well-known region in mountain research. The fairly small number of papers presented during the Oxford conference concerning this area confirms this feeling. Moreover, the various specificities of the mountains of this region prevent scientists from simply extrapolating to this context their understanding of dynamics measured elsewhere, in mountains which have been studied more. The mountains of Southern Europe, North Africa and the Near East are not among the highest on Earth; if several summits are more than 3000 m high, very few are higher than 4000 m. But, since most of them – volcanoes as well as limestone and crystalline structures – are fairly close to the sea because of an incredibly indented coast, slopes encircling the maritime basin are often steep, producing impressive landscapes and a huge variety of biotopes. This topographical context combined with a specific climate, known for its frequent extremes, its summer drought and very intense rainstorms, cause an average loss of soil through water erosion which is among the greatest in the world.

This natural fragility of the mountains has to be related to their traditional and modern human uses. The Mediterranean basin is an area of very ancient human settlement; agricultural practices, stock raising, intense exploitation of forests have taken place over more than two millennia. The coastline has been urbanised for about the same span of time, and long-lasting city-countryside relationships have been able to shape the landscape and the collective perceptions of what the mountain is or should be.

This combination of characters produces the feeling that this is a very specific mountainous area. Most of the papers presented here stress on one or several of these characters in a specific geographical context: Moore *et al.* for the Rif and Coelho *et al.* for north and central Portugal are studying the impact of cultivation and agricultural change on soil erosion; Coelho *et al.* stress the hydrological effects of changes on forest use in the same area. Two sets of authors underline the issues of economic changes on natural and social characteristics of the Greek mountains: Katsaros *et al.* for animal breeding and Efstratoglou *et al.* for rural tourism. Last, Funnell and Parish adopt a complex and global approach of local cultures of the High Atlas subjected to dramatic changes in their everyday life. Altogether, these papers, whose presentation was followed by an informal but valuable discussion, lead to a collective reflection about crucial issues: factors of land degradation; natural hazards, especially those related to hydraulic regimes; social and cultural changes; deforestation-reforestation processes.

However, these papers and the discussion that followed also lead to another kind of conclusion: everyone had to recognise the variety of local contexts and the difficulty of defining common Mediterranean characteristics that would be relevant everywhere around and close to the basin. A deep economic and cultural border between North and South crosses the Sea. Southward, the densification of rural settlements is real; the need for commercial agriculture, sometimes illegal ("Kif in the Rif"), is strong; the pressure on the natural environment is important. Northward, depopulation has been important during the last century, threatening the identity and stability of local communities; many traditional activities have weakened or disappeared, and this kind of change in human pressure on the natural environment is also causing regional and ecological problems. This north-south differentiation is reinforced by a biogeographical boundary between deforestation and reforestation, involving the ability of the ecosystems to react to contrasting changes in land-use. This boundary, however, seems to be located much further north than the first one. Last, most research undertaken in the Near East is not well-known enough in Europe to be taken into account by European scientists. This variety of mountains in these areas is well-known but, up to now, there have been too few efforts or results in comparative studies and research networking to allow scientists to obtain a comprehensive view of what simultaneously happens at international, regional and local scales around the basin. The relative scarcity of active comparative and international programmes is not going to improve our knowledge in the near future.

C.O.A. Coelho[1], A.J.D. Ferreira[1], R.P.D. Walsh[2] and R.A. Shakesby[2]
[1] Departamento de Ambiente e Ordenamento, Universidade de Aveiro
3810 Aveiro, Portugal.
[2] Department of Geography, University of Wales Swansea, Singleton Park
SA2 8PP, Swansea, Wales, United Kingdom

Hydrological responses to forest land-use in the central Portuguese coastal mountains

The coastal mountain range of central Portugal experiences a Mediterranean/ Atlantic transitional climate with a highly variable annual rainfall: over the past 60 years, the amount in the highest year has been 3.5 times the amount of the lowest. This can pose problems in terms of water management, namely with regard to water balances, water availability and peak magnitudes under extreme rainfall events. An extra level of variability results from the different land-uses that arise directly or indirectly from human activities. The mainly forested areas have undergone major land-use changes in the last two decades from *Pinus pinaster* Aiton to *Eucalyptus globulus* Labill., a process hastened by frequent and widespread forest fires.

This paper analyses the impact of different forest management types on: (1) the hydrological balance; (2) the hydrological response to rainfall events of various magnitudes at different monitoring scales, and particularly catastrophic events leading to damaging floods on downstream populations, and (3) water availability in dry periods.

To assess the human impacts on the water resources, three monitoring scales were adopted: (1) small bounded plots (8 m × 2 m) connected to wash traps and runoff measurement devices; (2) small (<1.2 km^2) catchments with dominant land-uses, allowing catchment water balances to be calculated; and (3) data from water level recorders of two small rivers with catchments $\cong$ 100 km^2 (Alfusqueiro and Água river).

Data are compared for three contrasting hydrological years (1989–90, 1990–91 and 1991–92). Values for the unburned mature-mix catchment are very instructive. Evapotranspiration is highest (1359 mm) in the wettest and least (954 mm) in the driest of the three years, but when expressed as a percentage of rainfall the reverse is true, with 73.7% in the wettest, but 96.8% in the driest year lost by evapotranspiration. The contrast in streamflow between the wettest and driest years is consequently very large. Whereas runoff totalled 484 mm in 1990–91, it amounted to only 32 mm in the dry year. A broadly similar pattern was true for the other unburned mature forest catchment.

The data from the remaining catchments reflect not only the contrasts in climate of the three years, but also the impact of post-fire vegetational recovery during the three-year period (3–6 years after fire) and contrasts in post-fire land management. In all three post-1986 fire catchments, runoff was higher and evapotranspiration (in both mm and %) was lower than comparative values for the unburned catchments. Evapotranspiration was highest, however, in the stump re-growth eucalyptus catchment, and least in the catchment where a large part had been rip-ploughed and planted with eucalyptus seedlings in 1988–89.

The main implication of the results is the sensitivity of the components of the hydrological cycle, and particularly streamflow, to changes in annual rainfall in the region. Even a modest decline in rainfall would result in a serious decline in streamflow. Thus, annual runoff in the unburned mature forest catchment in the dry year of 1991–92 was only 7% (32 mm) of the total in the near-normal rainfall year of 1990–91 (484 mm). In other words, a 40% decline in rainfall resulted in a 93% decline in streamflow. The water balance for 1989–90 suggests that a 20% rainfall decline would result in a 33% lower streamflow than in the near-normal year.

In respect to the streamflow response during the dry summer periods, a striking aspect is the absence of stream runoff for the three measured summer periods in the mature eucalyptus catchment. This is connected, at least in part, with the high eucalyptus evapotranspiration rates.

Depressional rainfall of low to moderate intensity, typical of temperate maritime areas, is supplemented by intense rainstorms characteristic of Mediterranean areas, which can cause catastrophic flooding. The land-use changes caused important changes in overland flow characteristics and the hydrological responses of both small and large catchments, with the highest flood production coming from disturbed forest areas.

Data are presented for a catastrophic event during Christmas 1995, with up to 300–400 mm in one week. A major feature was 10% overland flow on the mature pine plot. Overland flow at this plot had never recorded more than 0.5% in previous rainfall events over seven years.

At the catchment scale, the disturbed land-uses showed the highest responses, in both time and amount. In the pine re-growth catchment, peak flow reached as much as 20 mm/hr. This may reflect in part a shallow soil (often less than 20 cm deep) leading to widespread saturation of soils and hence overland flow, but also a well-developed macropore network (Ferreira, 1996), promoting rapid throughflow. The rip-ploughed eucalyptus comes second, followed by the eucalyptus re-growth catchment. The large catchments produced responses intermediate in size compared with the smaller response of the mature mixed-forest catchment and the post-1986 fire-disturbed catchments. One of the most interesting features is the failure of the mature forest catchment to respond until the second phase of the storm; this probably reflects the higher storage capacity of its soil.

REFERENCE

Ferreira, A.J.D. (1996). Processos hidrológicos e hidroquímicos em povoamentos de *Eucalyptus globulus* Labill. e *Pinus pinaster* Aiton. PhD Thesis on Applied Environmental Sciences, Universidade de Aveiro, p. 418.

C.O.A. Coelho, T.M.M. Carvalho and A.J.D. Ferreira
Departamento de Ambiente e Ordenamento
Universidade de Aveiro, 3810 Aveiro
Portugal

Land-use changes in Portuguese mountain areas: past, present and future

Mountain areas cover an important part of northern and central Portugal, supporting a lifestyle very distinct from those of the surrounding plateaux and plains. From the beginning of the century, this way of life, based on multi-purpose activities, became less attractive, giving rise to depopulation. Some areas have reached the threshold of abandonment of a permanent human presence in the mountains. It is now necessary to examine the present situation and to make proposals for measures to reverse this tendency.

This paper aims to discuss the preservation of a mountain lifestyle in the north and centre of Portugal, focusing on past evolution, the present status and the future tendencies of two representative mountain areas: Lousã and Caramulo mountains. These cases offer good examples of the constraints posed by the mountains in the light of population dynamics and economic development. The methodologies used were: (1) interviews with key actors such as local inhabitants, local authorities, foresters, development associations and producers' associations; (2) surveys of school populations; (3) analysis of demographic data and (4) assessment of land-use changes, through aerial photography interpretation and field work.

At the beginning of this century, there was a more or less uniform land-use throughout the mountain areas in this part of Portugal, based on multiple-purpose activities that included (1) subsistence farming; (2) grazing; (3) chestnut groves; (4) production of charcoal and wood for fuel; (5) small woods for timber. Some changes took place in the early decades when the fungal ink disease killed most of the chestnut groves and the central government decided to afforest the communal lands previously used for the grazing of sheep and goats, with *Pinus pinaster*. This led to emigration towards the littoral and to some European countries, particularly from the Lousã mountains, where most of the villages were abandoned.

The Caramulo mountains, being closer to a littoral industrial area, and more accessible, benefited from the installation in the surrounding plains of a pulp factory that required large amounts of wood, inducing a shift from *Pinus* to a faster-growing species, *Eucalyptus globulus*. As a result, a significant population was retained. Nevertheless, the two mountain areas have a high percentage of ageing population and a low proportion of young people, presenting serious problems of generation replacement since most of the latter do not want to remain in the mountains.

At present, the land-uses in the Lousã mountains are more diversified, with grazing, subsistence farming and forestry, especially pine and oak stands. In the Caramulo mountains, forestry is the dominant land-use, eucalyptus being the dominant species. This is mainly due to differences in land ownership. The privately-owned land in the Caramulo mountains allowed the planting of fast-growing profitable species, leading to a eucalyptus monoculture. In the Lousã mountains, since the 1920s, most of the land has been owned and managed by the government forest services, with resulting improved biodiversity.

However, the limited range of economic activities and the absence of employment opportunities in the secondary and tertiary sectors is less attractive to the remaining population, offering only work in traditional activities. The improvement of these activities has not always been successful, due to: (1) reduced cooperation among regional and local producers; (2) their reluctance to innovate; (3) the use of ancient techniques with low productivity; (4) reduced surface of pastureland; (5) low productivity of local goats; and finally (6) national and European Union laws that were not adaptable to local conditions.

In the Caramulo mountains, the environmental conditions introduced by the eucalyptus stands tend to be unsustainable due to impacts on soil erosion and water resources. The monoculture system also brings problems of a higher vulnerability to diseases and pests; forest fire risk; and especially the fluctuation of prices for paper pulp on international markets.

The Portuguese mountains were previously heavily populated, and traditionally man had promoted ecosystem stability. With their abandonment, factors such as forest fires and rip-ploughing for afforestation, among others, can disrupt this equilibrium. The presence of man in the mountain areas has an important role for the safeguard of natural resources.

The development of integrated and sustainable management of the mountains is necessary in order to reverse the desertion process and to rehabilitate mountain area activities, through the:

- Attraction of active populations;
- Maintenance or improvement of landscape diversity through multiple uses for rural development: (i) Sustainable forestry for conservation, fire prevention and diversity, (ii) grazing, (iii) biological farming, (iv) recreation and rural tourism;
- Promotion of traditional products and maintenance of the cultural heritage.

Jordi Corominas
Universitat Politècnica de Catalunya, Departamento de Ingeniería del Terreno
E.T.S. Ingenieros de Caminos de Barcelona
Spain

Management of natural hazards in Mediterranean mountainous regions

Extreme natural events are frequent in mountainous regions. Traditional societies were used to the occurrence of natural hazards, and their effects on individual, and largely self-sufficient, communities were of only local extent. In recent decades, however, many mountainous areas of the developed Mediterranean countries have experienced rapid growth. Urban settlements, communications and facilities have to share scarce space in the valley floors, often exposed to floods and landslides, and consequently more people are exposed to the hazards. These hazards may result in loss of human lives, the interruption of communications and the isolation of valley communities with significant losses to their economies.

The management of natural hazards in Mediterranean mountains is reviewed here. Hazard assessment requires an adequate knowledge of the mechanisms governing the natural systems, the nature of the phenomenon involved, its areal extent and its frequency. In the Eastern Pyrenees, for the most common hydrologically-related hazards (floods and landslides), two rainfall patterns have been identified: (a) very intense, short duration rains (above 200 mm in 24 hrs) cause flash floods and trigger first-time landslides, mostly rock falls, debris flows and debris slides; (b) rains of moderate intensity, lasting for several weeks, cause minor flooding and reactivate landslides, mostly rotational slides and mudslides on clayey formations. In this region, there has been an increase in the reactivation of landslides and in the frequency of floods since 1960. There are not enough data to determine whether this increase is due to climate change or to natural cyclic variability. However, given the importance that a change in the occurrence of natural hazards might have on the economy of mountain communities, this topic needs further research.

The concept of sustainable development should be applied to hazard management. Proper management should consider the natural systems globally, and the preventive actions undertaken should neither aggravate the consequences of the natural hazards nor transfer them to other locations. This is particularly relevant in flood control, in which actions undertaken at any point of the basin may force undesired changes in the river dynamics in both upstream and downstream reaches. Care should be taken to avoid the excessive use of structural measures, such as channeling, for flood control purposes. Often the role of the flood plain as an instrument for flood control is disregarded. In the Mediterranean regions, where flash floods are frequent, peak discharges last for only a few hours and transport a small volume of water. In such circumstances, water overflowing the river channel to be retained in the flood plain may reduce significantly the peak discharge in downstream sections, avoiding further damage. Channeling of rivers prevents bank overspill and transfers the peak flood discharge downstream. A similar effect can be expected from the invasion of the floodplains by constructions, such as road or railway embankments, acting as artificial levees.

Structural measures are expensive and may provide a false sense of security that favours further development of threatened areas, thus increasing the potential damage; by using structural works, risk can only be minimized, not eliminated. Engineering works for flood control are usually designed for events with long return periods, e.g., 500 years, and not for the maximum expected discharge. On the other hand, mountainous regions are fragile environments. Structural works can degrade the fluvial ecosystem. The construction of large dams for flood control causes changes in water quality, nutrient depletion, and the retention of the sediment necessary to maintain the replenishment of beaches and deltas. It is thus necessary to include non-structural measures for hazard management and, among these, land-use planning has proved to be the most efficient.

Sofia Efstratoglou[1] and Demetrios Psaltopoulos[2]
[1]Professor and [2]Research Fellow, Agricultural University of Athens
Department of Agricultural Economics, Iera odos 75, Athens 11855
Greece

The economic role of rural tourism in Greek mountain areas: the case of Evrytania

Rural areas in the European Union (EU) are currently undergoing significant economic and social changes, mostly induced by agricultural policy reform, international trade liberalisation, and the strengthening of the role of rural development policies. In particular, 'peripheral' rural areas – a significant part of which are mountainous – which are characterised by remoteness, depopulation, infrastructural inadequacies and high dependence on the agricultural sector are expected to be significantly affected by the above-mentioned developments.

Greek mountainous areas are part of the 'peripheral' rural areas of the EU. Agricultural adjustment in these areas is expected to reduce employment opportunities in farming, increase the dependence of local economies on welfare transfers and considerably threaten their fragile social fabric. Within this context, the development of economic activities outside agriculture seems to be the only answer to their socio-economic survival. Furthermore, the high recreational value of Greek mountain areas leads to the recognition that the expansion of tourism can promote economic development.

The objective of this paper is to investigate the economic role of tourism in the Greek mountainous region of Evrytania, through the application of Input–Output (I-O) Analysis. It is written in the context of an EU-funded FAIR3 Research Project on 'Structural Policy Effects in Remote Rural Areas Lagging behind in Development'.

Evrytania is inhabited by 24,000 people and has historically been one of the poorest regions in Greece, characterised by severe depopulation. Most of the comparatively elderly labour force is occupied in agriculture, and there are high rates of underemployment. The natural beauty of Evrytania (natural forest and substantial water resources) has recently led to the steady expansion of the tourism industry as national and EU funds, channelled towards infrastructure and agrotourism, have substantially improved access and accommodation facilities.

After the 1988 reform of the EU Structural Funds, and more specifically in the 1989–1993 period (First Community Support Framework), the escalation of the Integrated Mediterranean Programmes (IMP) and the introduction of the Community Support Framework (CSF) Regional Operational Programmes and the Agriculture National Operational Programme, as well as of Community Initiatives has led to the provision of over 22,000 million Drachma for structural policy investment in Evrytania (around 25% of the area's annual average gross domestic product in that period). Moreover, if we include Common Agricultural Policy (CAP) Guarantee spending during the same period (2,062 million Dr.), the increase of annual average spending for structural and agricultural policies is even higher. Almost 40% of total spending was allocated to road improvement and construction, followed by agriculture (21.5%), which significantly influences the survival of numerous communities, tourism infrastructure and development (14%), and forestry (5.7%).

From the above, it is obvious that the expansion of rural tourism, via the development of infrastructure and tourism facilities, as well as the sustainable management of the area's agricultural and forest resources, is considered crucial to Evrytania's economic development strategy. Consequently, we investigated the economic role of tourism via the application of the Generation of Regional Input–Output Tables (GRIT) technique to the estimation of a 1988 Input–Output table for the economy of Evrytania. The regional table was constructed in two steps: first, the estimation of the mechanically-derived input purchasing and output sales patterns of economic sectors (via employment ratios and location quotients) and, second, their modification through the insertion of superior data, derived from a business survey of 11 sectors (agriculture, forestry, fishing, food processing, textiles, timber processing, furniture, other manufacturing, construction, trade, hotels and catering). The criteria for the selection of those sectors for survey were firstly their importance to the local economy, and secondly their influence by structural policies.

Multiplier analysis showed that the tourism-related sectors of recreational services, hotels and catering and trade can generate significant economic effects in Evrytania. Recreational services generate significant output, income and employment effects, hotels and catering create an important number of jobs, while the indirect and induced effects of trade are also important.

As a next step, an impact analysis was carried out, trying to estimate the economy-wide impacts of tourism in Evrytania in 1988, via the 'disappearance' of tourism from the local economy in this year. In doing so, based on estimates of the business survey, final demand for the outputs of recreational services, hotels and catering and trade was reduced by the tourism-relevant expenditure amounts. Results showed that the impact of the 'disappearance' of the tourism sector in Evrytania would be the reduction of output by almost 9%, incomes would decrease by 13.6%, while 15.4% of jobs would be lost. The loss of economic activity would be severe in the case of hotels and catering, recreational services, trade, energy and fishing, while an important number of jobs are also lost in food processing.

Finally, we attempted to estimate, using standard Leontief procedures, the impact of increases in investment specific to the tourism sector (hotel and agrotourism establishment-building and improvements) for 1989 to 1993. After taking account of inflation, tourism-related investment was found to amount to 1,859.37 million Dr. (1988 prices). This amount represented demand for the output of the construction sector. Tourism-related investment activity increased local output by 8.3%, incomes by 14.1% and employment by 15%. Changes in output were very significant in construction, while there were important increases of incomes in timber processing and trade, and of employment in food processing, timber processing, trade and transport.

In conclusion, mountain areas in Greece have been traditionally faced with significant structural constraints such as remoteness, lack of accessibility, severe depopulation, lack of infrastructure and dependence on agriculture. At the same time, certain features that have caused their isolation have contributed to their status as areas of natural beauty, unspoilt environment and rich cultural heritage. In these areas, rural tourism and related activities could be considered as the main axis of a rural development strategy that can protect the fragile socio-economic fabric in the short run, and create sustainable development in the long run. The economic importance of rural tourism in the case of Evrytania, appearing through the I-O multiplier analysis and the estimation of related economy-wide output, income and employment impacts, provides strong empirical evidence for the above statement.

Don C. Funnell[1] and Romola Parish[2]
[1]University of Sussex, AFRAS Arts C, Falmer
Brighton BN1 9QN, United Kingdom
[2]Department of Geography, University of St Andrews
St Andrews, KY16 9ST, United Kingdom

Complexity, cultural theory and strategies for intervention in the High Atlas of Morocco

Introduction

Price and Thompson (1997) argue for the use of Cultural Theory in the analysis of mountain communities, whilst Durand (1997) echoes earlier work in Switzerland using ideas of dynamic modelling in the understanding of the transformation of mountain economies. This paper argues that the notion of complexity (Clark *et al.*, 1995) underlies these ideas and examines their applicability to the High Atlas region of Morocco.

The High Atlas

Several studies (Parish and Funnell, 1996; Messelli, 1996; Bellaoui, 1996) provide the basis for a discussion on land degradation, the introduction of new tree crops, pressures on grazing regimes and new tourist initiatives in the High Atlas. The literature on each of these problems reveals distinct sets of theoretical approaches and subsequent policy initiatives. There is often little or no recognition that each of these issues is but one facet of the total livelihood of the mountain communities, and there appears to be no easy way of assimilating them all within a broad theoretical framework that provides a useful explanation for patterns of behaviour.

Complexity and social theory

Work by biologists, physicists and, now increasingly, some social scientists is drawing on the notion of complexity to explore problems that are not amenable to typical reductionist models.

Most writers emphasise that complex systems are indeterminate in the long run. They are not simply self-replicating structures which continue *ad infinitum*. Therefore the nature of their evolution is central to understanding long-term change, as there must be 'mechanisms' which affect the trajectory of change. Thompson (1997) has highlighted the problem of scale, and the simultaneous evolutionary dynamics at different scales add a further element of complexity to the system. The result is that attempts to provide policy prescriptions (especially those labelled sustainable) require a different framework from that usually suggested.

Cultural Theory

Some of the key ideas presented above find resonance in the literature of behaviourist social theory. Thompson *et al.* (1990) argue that society and individuals hold a number of generalised ideas or 'myths' about the way nature behaves, and from these myths are derived several categories of social or organisational behaviour. Nature may be considered benign, or as proving clear limits to use, or as extremely sensitive, or completely capricious. Attitudes to management vary according to the perspective adopted. Individuals or groups alter their viewpoints over time and for different circumstances. Faced with a novel situation, a community might adopt any one of these myths and there is no certainty that a particular outcome will arise.

Complexity and Cultural Theory in the High Atlas

Whilst, at one level, these ideas may be little more than re-labelling existing theoretical perspectives, such criticism fails to capture some of the deeper issues.

1. Co-evolution and scale

The problem of degradation is re-examined, pointing to the lack of research exploring the decision processes of land-use and reactions to soil erosion. The changing landscape and farmer responses need to be examined at different scales. Work under the MEDALUS programme (Clark *et al.*, 1995) is noted, which looks at a cascade-type dynamic model in which different processes are triggered at varying spatial scales. This work, carried out within the framework of complexity theory, has shown that we cannot be certain that the processes highlighted at one scale will have a predictable (and therefore straightforwardly manageable) outcome at another scale. Therefore policy prescriptions need to develop much greater sensitivity to uncertainty and scale.

2. Complex systems feed upon diversity

The value of cultural diversity as a policy objective is not yet fully appreciated. Culture, in this sense, concerns the panoply of behavioural norms and supporting institutions that sustain communities. The Berber communities of the High Atlas have developed a cultural framework which has absorbed changes in both the variety of production strategies and their social organisation. Much of the 'policy orientated' research assumes that there is only one answer to any set of problems, usually through an extension of the formal bureaucracy. The process of development is leading to a reduction in the institutional variety within the mountains.

3. Complexity and indeterminacy

Both theories argue that we cannot 'know the future' and suggest that we must build change into our understanding and interventions in mountain communities. Thus 'keeping options open' may well be the best survival strategy. Unfortunately, the trends in the economy are prioritising certain activities, for instance production of specific cash crops and, to some extent, tourism. Simply 'protecting existing institutions' is insufficient unless these organisations possess the ability to metamorphose into new structures to cope with the changing world. Wider incorporation of the communities has meant that there has been a gradual transference of the 'security' responsibility outside the local community.

Conclusions

The problems facing mountain communities need to be tackled through a recognition of the 'complexity' and the existence of multiple 'cultures'. This has important implications for both research and policy. Not only should we abandon a predilection for 'simple explanations' but recognise the importance of the mechanisms that allow communities to regulate change and to learn from this experience how different responses may co-exist within one community. Indeterminacy is an inherent feature of social change which argues for flexibility. Maintaining flexibility is not only a technical agenda but also political, invoking ideas of subsidiarity.

REFERENCES

Bellaoui, A. (1996) Tourisme et développement local dans le Haut-Atlas marocain: questionnement et résponses. *Revue de Géographie Alpine*, **84** (4): 15–25.

Clark, N., Perez-Trejo, F and Allen, P. (1995) *Evolutionary Dynamics and Sustainable Development*. Edward Elgar, Aldershot.

Durand, M-G. (1997). Un "système montagne" réinventé: le développement durable dans les Alpes françaises ou la nouvelle gestion globale des territoires. *Revue de Géographie Alpine*, **85** (2): 157–172.

Messelli, D.A. (1996) Constraintes d'une utilisation durable des ressources naturelles du Haut-Atlas: le cas du bassin intramontagnard de Tagoundaft. *Revue de Géographie Alpine*, **84** (4): 109–119.

Price, M.F. and Thompson, M. (1997) The complex life: human land-uses in mountain ecosystems. *Global Ecology and Biogeography Letters*, **6** (1): 77–90.

Parish, R and Funnell, D. (1996) Land, water and irrigation and development in the High and Anti Atlas mountains of Morocco. *Geography*, **81** (2): 142–154

Thompson, M., Ellis, R.J and Wildavsky, A.B. (1990) *Cultural Theory*. Westview Press, Boulder, Colorado.

Thompson, M. (1997) Security and Solidarity: an anti-reductionist framework for thinking about the relationships between us and the rest of nature. *Geographical Journal*, **163** (2): 141–149.

Dimitris Katsaros
National Agricultural Research Foundation
Institute of Mountain and Rural Economics
2nd Km. National road Karpenissi â Lamia
36100 Karpenissi
Greece

Animal breeding in the mountains of Greece: encouraging perspectives for the development of small family farm holders

Understanding the development of mountain stock breeding

Extensive mountain stock breeding in Greece has succeeded not only in surviving but also in employing a large part of the population. This is even since the introduction of State boundaries and limits that significantly limited the free passage of the animals from one mountain range to another, and prohibitions on entering what was supposed to be state-forest land, as well as subsidy and level compensation policies.

We must now be more realistic and understand animal breeding at the local level, discovering local potentials according to the conditions that can be investigated, using existing perspectives, of small family-owned stock breeding operations in the mountains. Mountain areas are currently going through social and economic changes because of experimental or pilot programmes suggested and implemented by the planners of both Greek and European development programmes.

The philosophy behind those programmes, being used for the first time in many mountainous areas, is on internal or self-development. In other words, these programmes are based on the activation of human labour and offer radical solutions for the regeneration of local economies. These refer mainly to the expansion of the tertiary sector – mainly rural tourism – but also to the secondary sector, with an emphasis on establishing and running small handicrafts and other businesses.

The passage from traditional production systems to new economic activities in traditional societies is a milestone for the development of mountain regions which have been neglected for years, and contributes to new conditions for including mountain economies in the general economic system of the country. However, there is a tension between the programmes and the traditional societies and economies that are affected. The dynamics of this tension are so important that they should be investigated before the investigation of radical changes offered by European and national organisations through such programmes; particularly Integrated Mediterranean Programmes (IMP) and LEADER programmes.

Whatever one's reservations regarding the programmes, mountain stock breeders must succeed in the local market, given the problems associated with selling to distant city markets. This opportunity has to be realised as, recognising the tough competition in breeding products, selling in the local market is the only competitive advantage offered to breeders.

From such an understanding, one can come closer to solving the financial problems of mountain breeders. Money paid to feed the animals can be found from the higher selling prices of quality-assured breeding products. Yet, for this to become a reality, consistent and continued education of breeders is needed. Significantly increased levels of development in various types of tourism provide an opportunity for animal breeding to succeed in the new market conditions.

Rural tourism and livestock

As noted above, an important opportunity for the development of breeding is entrance to local markets which have been strengthened by the introduction of new economic models for local development, resulting in a continually increasing number of visitors to rural areas. One example is provided by the prefecture of Evritania, in central Greece, a region that is entirely mountainous. After a continuous decrease (3–4%/year) in numbers of livestock from the early 1970s, numbers have stabilised, with a slight increase in the number of locally bred small ruminants in the past five years. In the same period, tourism has increased significantly. Without doubt, there is a relation between the stabilisation and increase in livestock numbers and the growth of tourism.

The growth of tourism, especially alternative types – particularly rural or agrotourism – provides a very good opportunity for stock breeding to succeed in new market conditions. Indeed, in Evritania and other rural mountain areas of Greece, the harsh climate that limited development and made the survival of the population even more difficult, is today an almost necessary condition for its wellbeing. The reason is that, since about the mid-1980s, winter sports, especially skiing, and also a desire for mountain experiences in general, have become ever more popular with city dwellers.

Since the Integrated Mediterranean Programmes in Rural Evritania began in 1990, approximately 85% of investment has been used for agrotourist lodgings. Until the end of 1992, 335 first-class rooms (about 800 beds) were built. Under the LEADER I programme, the number was increased by 50 rooms (120 beds), and a further increase of 100 rooms is forecast under LEADER II. Evritania now has a capacity of 1,050 beds merely in agrotourist lodgings. There are also municipal guesthouses and hotels. In 1996, the total number of beds was 2,450; quite considerable for a prefecture with 11 inhabitants/km^2, 300 km from Athens, and 80 km from the largest city, Lamia.

The traditional Greek rural *œphiloxenia* (hospitality) is based strongly on the consumption of lamb and goat during spring and summer and pork in winter. Consequently, the local demand for meat and meat products (cheese, Greek feta and butter) increases continually, and the points of sale (i.e., butcher shops) in the prefecture increased by 60% from 1990 to 1995: from 25 to 42. Of these, only eight are in Karpenisi, the prefectural capital, with 10,000 inhabitants. The remaining 34 butcher shops are scattered between the villages of Evritania, which have an average population of 150: far too low a number to justify their existence simply for the needs of the native population.

Basic suggestions

Given the trends mentioned above, activities supporting mountain stock breeding are urgently needed, before the scale of renting rooms reaches the catastrophic levels of the islands. Entrance to the local market should be eased for mountain stock breeders. This must be done sensitively because major changes will influence the rural economy and society in the near future. The following suggestions do not aim to give an overall picture for the development of stock breeding. However, they do aim to awaken the sensitivity of the responsible services for immediate and low-cost actions towards supporting stock breeders:

- the state and local developmental agencies should focus their attention on educating the stock breeders, many of whom ignore existing capabilities and cannot find a way to enter larger markets;
- the scientific community should also direct their attention to research at the national level on products that can be classed as natural, pure, or should receive urgently-needed labels of origin;
- local community leaders should co-operate in order to start programmes to encourage stock breeders, increase their level of co-operation, and improve the flow of information to a more direct level.

REFERENCES

Anthopoulou T., Gousios D., Katsaros D. (1988) Social and economic evolution of the Greek rural land. Problems and the future potentials: The example of Evritania in the disadvantageous Mediterranean zones in EC-INRA, CEMAGREF-IAM, Vol. 2, pp. 245–275.

Giannakopoulos A., Tassos G. (1993) Conclusive developmental plan for stock breeding in Evritania prefecture. EVRITANIA S.A., Rural Development Agency.

Katsaros D., Anthopoulou T. (1987) Allocation and new economic activities in Evritania. *Bulletin de la Société Languedocienne de Géographie*, **3–4**: 329–344.

Katsaros D. (1992) Self-developmental dynamics in mountainous rural land: some local examples in the Agriculture in the 1990s: economical and social perspective. Panhellenic Convention on Rural Economy, pp. 249–260.

Katsaros D. (1997) The Mountain Livestock Systems International Conference of the ELPEN Network. Karpenissi, 22–24 January.

Heather Moore[1]. Howard R. Fox[1], Cherif Harrouni[2] and Ahmed El Alami[2]
[1]Centre for Land Evaluation & Management, Geography Division
University of Derby, Kedleston Road, Derby DE22 1GB, United Kingdom
[2]Institut Agronomique et Vétérinaire Hassan II, Complexe Horticole, Agadir, Morocco

"Kif in the Rif": biodiversity under threat in northern Morocco

The Rif environment

The Rif Mountains comprise heavily folded and faulted sedimentary and metamorphic rocks rising to over 2400 m. in a narrow and deeply dissected mountain range. The mountains form a barrier to the main climatic systems moving in from the Atlantic, and in consequence precipitation totals decline from 1400 mm in the western and central parts of the Rif to 300 mm on the Mediterranean coast and in the east. In places, soils are relatively deep and fertile, but where slopes are steep, soils are thinner. Cedar and oak species dominate the vegetation of the higher more humid areas, with Thuya and maquis typical of the drier regions.

The agro-sylvo-pastoral system of the Rif region is adapted to the opportunities and limitations of the mountain environment. In Chefchaouen Province, only 20% of the land area can be cultivated because of the dissected nature of relief and shallow soils. On the cultivated land, cereal production, market gardening and arboriculture dominate, with wheat occupying 30,000 ha and olives 28,000 ha. Livestock rearing is a traditional and important part of the agricultural system, with cattle, sheep and horses in the area of high agricultural activity, and goats grazed in the higher mountain rangelands. In Chefchaouen Province, 90% of the population is rural and almost 80% employed in agriculture (DPA, 1994).

The Rif Mountains have been identified as an important 'hot spot' of biodiversity within the Mediterranean basin with over 2,000 species per 15,000 km^2 and endemism of greater than 20% (Médail and Quézel, 1997). Among the endemic species, the Moroccan fir *Abies maroccana* is perhaps the most impressive example, but several Eurosiberian linking species and wild progenitors of cultivated crops and food legumes can also be found in the region (FAO, 1996).

The problem

The biodiversity of this region and indeed the agricultural economy is under severe threat from landscape degradation. Forest clearance over many centuries has greatly reduced natural woodland cover, a process that is continuing.

Vegetation clearance also leads to the acceleration of soil erosion and hence further degradation. It is estimated that 60% of all soil lost through water erosion in Morocco is derived from the Rif Mountains, with some of the highest specific sediment yields in the world (McNeill, 1992; Fox *et al.*, 1997).

Environmental degradation in the Rif Mountains is set against a background of rapid population growth in the region over the last 40 years. For example, in Chefchaouen Province, the density of population has risen from 40 persons/km^2 in 1971 to 101 persons/km^2 in 1994. These population changes have placed great pressure on the natural landscape, and have in part given rise to three of the principal causes of degradation:

- Non-sustainable use of forests: After centuries of unregulated forest clearance, forest management is now much tighter. Despite this, illegal and unregulated exploitation of natural forest leads to losses of between 1,000 and 1,500 ha. per year in Chefchaouen Province for example (DPA, 1994).

- Change in the agricultural system: Expansion of the cultivated area into marginal lands at high altitudes is symptomatic of the pressure placed on agriculture by an increasing population. Between 1966 and 1986, the cultivated area increased by 93% in the western central Rif (Boukil *et al.* 1987). Reduction in fallow period length, increasing mechanisation and the use of fertilisers and herbicides all pose threats to the natural ecology of the region.

- Cannabis cultivation: Cannabis, or kif, (*Cannabis sativa*) is a traditional crop of the region. However, in recent years, a dramatic growth in the area of cultivation has taken place and the farming of this crop represents a major threat to the Rif landscape. In 1996, the area under such cultivation was 74,000 ha; much of the expansion has occurred on less accessible, marginal land (USDS, 1996). Cannabis cultivation and trade now represents the largest source of hard currency for the Moroccan economy, equivalent to 60% of legal exports. In 1994, it was estimated that 200,000 people lived directly or indirectly from the cultivation of cannabis (Laouina, 1995). Table 1 shows the levels of income and profit that may be gained from kif in comparison to other crops.

Table 1: Comparative average annual investment, yield, income and profit for different crops grown in the Rif region

	Investment Dh/ ha	Yield t/ ha	Income Dh/kg	Income Dh/ha	Profit Dh/ha
Cereals	2000	2	2.3	4600	2600
Tomatoes	40 000	60	1	60 000	20 000
Potatoes	13 000	15–20	1	20 000	7000
Olives	10 000	10	2	20 000	10 000
Cannabis	6000	0.7	dried leaves 45 resin 7000	33 750 52 500	27 750 46 500

The way forward?

Over a number of years, attempts have been made to reduce the effects of environmental degradation and to provide solutions to the problems of the region. The main initiatives have included the following:

- A counternarcotics programme to attempt to curtail the trade in cannabis. This was begun in 1992 under the auspices of King Hassan II and funded jointly by European Union and the Moroccan Government;
- The creation of Talassemtane Natural Park covering 60,000 ha in Chefchaouen Province to protect the remaining areas of *Abies maroccana* and associated endemics (AEFCS, 1996);
- The development of research programmes concerned with the diversification of agricultural activities and a reduction in Cannabis cultivation. For example, plot trials of crop substitutes such as apples and pears are being conducted near Bab Taza;
- The establishment of a programme of 'désenclavement' involving the provision of improved infrastructure in the region, including road access, water and electricity.

Such measures have been proposed and adopted in a relatively piecemeal fashion over a number of years.

An integrated approach making use of GIS methodology, landscape ecology and gap analysis within the framework of the Biosphere Reserve concept (UNESCO, 1996) is more likely to produce successful strategies for the safeguarding of the natural environment of the region as well as the economic and social welfare of the people.

REFERENCES

AEFCS (1996) Etudes sur les Aires Protégées, Rapport de Synthèse. Royaume du Maroc, Rabat.

Boukil, A.F., Blali, A. and El Kassi, M. (1987) Evaluation et cartographie de la dégradation des forêts dans la zone nord. Ministère de l'Agriculture et de la Réforme Agraire. Centre Régional de l'inventaire et des aménagements, Tetouan.

Direction Provinciale de l'Agriculture (DPA) (1994) Potentialités et développement agricole dans la Province de Chefchaouen. Unpublished Internal Report Direction Provinciale de l'Agriculture, Chefchaouen.

FAO (1996) East and South Mediterranean Sub-Region Synthesis Report. The Fourth International Technical Conference on Plant Genetic Resources, Leipzig, Germany 17–23 June 1996. FAO ITCPGR/96/REP. Rome.

Fox, H.R., Moore, H.M., Newell Price, P. and El Kasri, M. (1997) Soil erosion and reservoir sedimentation in the High Atlas Mountains, southern Morocco. In Walling, D.E. and Probst, J.-L. (eds) *Human Impact on Erosion and Sedimentation*. International Association of Hydrological Sciences Publication 245: 233–240.

Laouina, A. (1995) Démographie et dégradation des sols dans le Rif. In De Noni, G., Roose, E., Nouvelot, J.F. and Veyret, Y. (eds) *Environnement Humain de l'Érosion*. *Réseau Erosion – Bulletin* (FRA), **15**: 69–77. Journées du Réseau d'érosion, 20–22.

Médail, F. and Quézel, P. (1997) Hot-spots analysis for conservation of plant biodiversity in the Mediterranean basin. *Annals of the Missouri Botanical Garden*, **84**, (1): 112–127.

McNeill, J.R. (1992) T*he Mountains of the Mediterranean World: An Environmental History*. Cambridge University Press, Cambridge.

UNESCO (1996) *Biosphere Reserves*. UNESCO, Paris.

USDS (1996) International narcotics control strategy report, March 1996. US Department of State. USIS, American Embassy, Stockholm.

Panagiotis Platis, Dimitrios Trakolis and Ioannis Meliadis
Forest Research Institute, Agricultural Research Foundation
570 06 Vassilika, Thessaloniki
Greece

Classification and evaluation of the rangelands of the Voras and Tzena Mountains of Northern Greece

Rangelands cover 65,440 ha of the Voras and Tzena mountains in northern Greece: a substantial percentage of the total area. The majority of the region is covered by natural forest ecosystems, as well as grasslands, shrublands and partially forested areas. The main land-use in the forest areas is forage production for farming and wild animals.

The inventory, classification and evaluation of rangelands in Voras and Tzena mountains was accomplished using orthophotomaps and Geographic Information Systems (GIS) through ENVIREG Programme concerning the recognition and evaluation of avifauna biotopes for inclusion in the European Union's Network for Regulation 79/409 (Platis et al., 1995; Trakolis et al., 1995).

The areas which were studied were: farms abandoned for more than five years; grasslands with woody species comprising no more than 10% of the plant cover; evergreen scerophyllous shrublands with shrubs no more than 5 m high; deciduous shrublands with shrubs no more than 5 m high; and partially forested areas with crown cover less than 40% and timber stock less than 100 m³/ha. The rangelands were classified according to the dominant vegetation (Wenger, 1984; Hunter and Paysen, 1986). Each inventory unit was classified into types, forms, series and subtypes of range vegetation and was evaluated for its surface area (Papanastasis, 1989). Afterwards, the orthophotomaps were digitized and all the relative information was recorded in a database. Several thematic maps were reproduced, exploiting the potential of GIS applications (Stone et al., 1994).

The information collected included surface areas, vegetation type, soil type, site class, range condition, bioclimate and altitudinal distribution. The inventory of the rangelands (grasslands and shrublands) showed that the largest proportion of the rangelands surveyed had four site classes and characteristics of good condition. They were distributed in the high elevation (801–1,200 m) and very high elevation (>1,200 m) zones. The grasslands (>1,200 m) of the Voras and Tzena mountains have the highest productivity of all the types. Therefore, rational management must be applied first to these areas. They cover 25% of the total area of 65,440 ha of the mountains, satisfying the needs of the grazing animals during the summer period; additionally they are considered as reserves for the fauna of the region. The small number of the animals, in comparison to the past, and the current grazing regime both play an important role for conserving the biodiversity of the region.

A higher percentage of the grazed area in the Voras and Tzena mountains, mainly in the sub-alpine zone, belongs to the second site class, with fair range condition. Most of the area of evergreen shrublands belongs to the second site class (65%) with fair range condition (58%), located in the intermediate and high elevation zones (80%) and on metamorphic rocks (50%). The majority of deciduous shrublands belong to the second site class (68%), with bad range condition (60%), found on metamorphic rocks (40%) and hard limestone (30%).

REFERENCES

Hunter, S.C. and Paysen, T.F. (1986) Vegetation classification system for California, user's guide. USDA Forest Service Pacific Southwest Forest and Range Experiment Station, General Technical Report PSW–94.

Papanastasis, V. (1989) Rangeland survey in Greece. *Herba*, **2**: 17–20.

Platis, P., Trakolis, D., Meliadis, I., Panagiotopoulou, M. and Tsougrakis, J. (1995) Recognition and evaluation of avifauna biotopes for inclusion in the E.U. Network of the Regulation 79/409. Biotope of Mount Pinovo-Tzena. ENVIREG Programme, Ministry of Environment Planning and Public Works. (In Greek)

Stone, T.A., Shlesinger, P., Houghton, R.A. and Woodwell, G.M. (1994) A map of the vegetation of South America based on satellite imagery. *Photogrammetric Engineering and Remote Sensing*, **61** (5): 541–551.

Trakolis, D., Platis, P., Meliadis, I., Tsougrakis, I. and Panagiotopoulou, M. (1995) Recognition and Evaluation of avifauna biotopes for inclusion in the E.U. Network of the Regulation 79/409. Biotope of Mount Voras. ENVIREG Programme, Ministry of Environment Planning and Public Works. (in Greek)

Wenger, K.E. (1984) Range management and ecology In: *Forestry Hand Book*, 2nd edition, pp. 739–800. John Wiley and Sons, New York.

Dimitrios Trakolis, Panagiotis Platis and Ioannis Meliadis
Forest Research Institute
570 06 Vassilika, Thessaloniki
Greece

The ecological importance of Mount Voras, northern Greece, and land-use changes in the prospect of sustainable development

Mount Voras, in northern Greece, has an area of 52,286 ha, with altitudes ranging from 200 m to 2,524 m. More than 95% of the area is covered by natural ecosystems, comprising forests and partially forested areas (62.2%), and grasslands and bushlands (32.9%); the remainder is agricultural land. In the foothills are 18 small communities with a total population of 15,460. Most of the area (80%), belongs to the state; the rest belongs to the communities or is private land. Since the liberation of this part of the country in 1912, the main land-use has been forestry, with sheep grazing during summer in the sub-alpine grasslands. The management authority is the Forest Service, and the exploitation of the forests, which produce 110,000 m³ of timber, is carried out by the local workers' co-operatives.

There is a great diversity of vegetation in the area, which in many places is at its final stage of succession. However, the presence of intermediate stages, with intensive phenomena of ecological succession, is very important. As a result of the large difference in altitudes, four vegetation zones (of the five found in Greece: Dafis, 1976; Moulopoulos, 1965) are present: the para-mediterranean zone (*Quercetalia pubescentis*), the beech-fir vegetation zone (*Fagetalia*), the zone of cold-resistant conifers (*Vaccinio-Picetalia*), and the zone above timberline (*Astragalo-Acantholimonetalia*).

The most common mammals of the area are *Lepus europeus*, *Vulpes vulpes*, *Canis lupus*, *Meles meles*, *Sus scrofa*, *Erinaceus europeus*, *Sciurus vulgaris* and *Martes foina*. Over the last five years, the population of *Capreolus capreolus* has increased. Other rare species are *Rupicapra rupicapra* and *Ursus arctos*. For the avifauna, seven habitats have been distinguished, based on the type and structure of vegetation, altitude, topography and past and recent human activities: (a) semi-mountainous habitat, (b) anthropogenic places in forests, (c) oak forest, (d) beech forest, (e) pine species habitat, (f) sub-alpine grasslands, and (g) rocks and gorges. From the ornithological point of view, the mountain is very important because of the presence of 119 bird species, of which 105 nest. Of these, 28 are referred to in annex I of European Union Directive 79/409, 39 are Species of European Conservation Concern (SPEC), and 15 are included in the Red Data Book of Endangered Vertebrates of Greece (Trakolis *et al.*, 1995).

There are four designated protected natural areas: two natural monuments and two permanent game reserves. The monuments have the status of strict nature reserves, and include a section of virgin forest of *Fagus sylvatica* and a mixed forest of conifers and broad-leaved species including *Pinus peuce*, which is found only in two places in Greece and it is strictly protected by special legislation only in this area.

Until ten years ago, the only people who had interests in the area were the local people employed by the Forest Service in various activities; the sheep farmers who have been using the sub-alpine grasslands for decades; the owners of the small agricultural farms in the foothills; hunters; those using or visiting the therapeutic baths at the edge of the area; and those who had leased from the state the four small mines for extracting marble, travertine and dolomite.

During the last decade there has been a shift from traditional land-uses (forestry and grazing) towards multi-purpose land-use, the final aim being the sustainable development of the area. To this end, two categories of studies have been undertaken recently. The first have been tourism-development-oriented studies, sponsored by the Ministry of Planning, Environment and Public Works and based on an interdisciplinary technical report of the Greek Tourism Organisation. The aim of these studies is to provide facilities for tourism development of the little-populated area around the mountain, based on activities such as skiing; mountaineering; forest recreation; controlled hunting and fishing; ecological, farm and therapeutic tourism; mountain biking; motor-cross and flying competitions. Secondly, nature-conservation-oriented studies, sponsored by the European Union under the ENVIREG and LIFE Programmes, have concluded that several parts of the mountain should be set aside and designated for protection.

The specific proposals from the latter studies include (a) the designation of three nature reserves: a peat bog for its ecological and palaeobotanical significance (Athanasiadis and Gerasimidis, 1986), a forest area with a great variety of flora and fauna species, geomorphologic formations, ornithological value and outstanding natural landscape, and two parts of forests of *Fagus sylvatica*, for their scientific, ecological, ornithological and aesthetic value; (b) the designation of a formation of caves as protected natural landscape, for its palaeontological, prehistoric and aesthetic value; (c) the establishment of a new game reserve; and (d) the inclusion of the whole area in the European Union Network of Special Protection Areas (SPA), due to its great ornithological importance, in accordance with the Directive 79/409.

The prospects of the area for sustainable development seem to be encouraging. Forestry, with selective cuttings leading to uneven aged stands, will continue to be the main activity on the mountain, recognising that the animal stock has stabilised for the last 35 years, and that sheep grazing will remain the dominant summer activity in the sub-alpine grasslands. Meanwhile, new interests have appeared in the area: tourism and nature conservation. Due to its ecological value, the area has been included in the EU LIFE Programme for the next three years. The timetable of actions to be carried under this programme includes management plans for the areas to be designated as protected; provision of environmental education facilities; and establishment of a monitoring system for many factors, such as evolution of the forest ecosystems, the dynamics of bird and mammal fauna, and the effects of grazing and tourism on the environment.

The area is accessible from many directions and is about 140 km from Thessaloniki, the second largest city in Greece. As it may also approached from the Former Yugoslav Republic of Macedonia, it is expected to become of international importance, not only for its ecological, aesthetic and recreational values, but also because it will become accessible from the north, since new entry points between Greece and the neighbouring country are anticipated.

REFERENCES

Athanasiadis, N. and Gerasimidis, A. (1986) Post-glacial evolution of the vegetation of Mount Voras. Annals of the Department of Forestry and Natural Environment, University of Thessaloniki, **29** (4): 212–249. (in Greek).

Dafis, S. (1976) Classification of forest vegetation of Greece. Ministry of Agriculture, special issue no. 36. (in Greek).

Moulopoulos, C. (1965) The beech forests of Greece. Annals of the Department of Agriculture and Forestry, University of Thessaloniki. (in Greek).

Trakolis, D., Platis, P., Meliadis, I., Tsougrakis, I. and Panagiotopoulou, M. (1995) Recognition and evaluation of avifauna biotopes for inclusion in the E.U. Network of the Regulation 79/409. Biotope of Mount Voras. ENVIREG Programme, Ministry of Environment Planning and Public Works. (in Greek).

Tor H. Aase
Institutt for Geografi, University of Bergen, Breiviken 2
5035 Bergen-Sandviken
Norway

Introduction: Western and Central Asian mountains

The mountains of Western and Central Asia are characterised by particularly dramatic transformations, in terms of both geomorphological processes and socio-political events. The region is tectonically very active; it is home to the greatest glaciers on Earth outside the polar areas; and its extreme altitudes and relative relief make it especially prone to diurnal erosion, landslides, and soil creep. From a social perspective, wars and conflicts are presently salient features, especially in the former Soviet Republics, in Afghanistan, and in Kashmir.

In such a politically and environmentally sensitive region, the notion of sustainable development is particularly challenging. All contributors to this section agree to the basic assumption that detailed knowledge of the interactions between natural processes and human activity is a precondition for sustainable resource management, favouring multidisciplinary, localised research. The chapter also testifies to the complexity of mountain geoecology systems, and to difficulties of generalising relationships between selected parameters. In the Pamir, Merzliakova and Sorokine relate land degradation to depopulation, while Zheng concludes the opposite on the Qinghai-Xizang Plateau (Tibet). Also, Aase argues that deforestation in Hindu Kush must be seen in relation to local politics, while the same phenomenon (deforestation) is usually placed in a context of farming and demography in Nepal.

As Western and Central Asia is one of the most inaccessible regions in the world, its mountain geoecology may erroneously be perceived as being an isolated issue. However, the links between mountains and surrounding lowlands are becoming increasingly intimate. Kreutzmann stresses the upstream-downstream aspect of hydrology in the Karakorum, integrating mountain societies with the plains below. The opposite link is demonstrated by Warner-Merl from the Altai, where mountain ecosystems are particularly vulnerable to the toxic air pollutants from industrial cities on the plains.

Although agreeing on most points, the following papers nevertheless reveal divergent opinions on one important issue: the relative importance assigned to local knowledge and external expertise in approaching questions of sustainable development. Kreutzmann favours the local experience accumulated through centuries in the management of water resources in the Karakorum. In line with his argument, Merzliakova & Sorokine demonstrate the failure of centralised planning in coping with agricultural production in Pamir during the Soviet period. In contrast, Zhu is confident that engineering and advanced technology are able to solve environmental problems in the southern parts of the Tibetan Plateau, where an unfortunate coincidence of great economic development potential and sensitive geomorphology occurs. Zheng also seems to argue in favour of central planning in order to solve problems of overgrazing in Southern Tibet. Probably, it is differences of epistemological positions between European and Chinese academic discourses that are surfacing here. In our quest for sustainable development, is it reasonable to adapt to a dramatically changing environment and listen to experienced local voices, or should we try to control nature, having confidence in science and engineering? This section offers no unambiguous answer to this question.

Tor H. Aase
Institutt for Geografi, University of Bergen
Breiviken 2, 5035 Bergen-Sandviken
Norway

Converting forests into weapons in the Western Himalayas

The issue to be addressed is that of the social context in which deforestation takes place. In this paper, it is argued that the context of deforestation is very different in the Upper Indus Northern Areas of Pakistan from that usually postulated by researchers in Central and Southern Himalaya, especially in Nepal. By implication, measures to check or control deforestation – if that is wanted – must also be dissimilar in the two regions. Strategies that have proved successful in Nepal cannot be transferred to Northern Pakistan, or vice versa.

In Nepal, the ongoing discourse on sustainable development locates the phenomenon of deforestation in the context of farming practices and demography. The core argument is well known: a reduced human mortality and a persistently high fertility rate have generated an increased demand for farmland in order to feed a growing population. Farmland is increased by constructing 'khet' and 'bari' terraces on marginal and sensitive land, or by extending slash and burn cultivation. Both means of expanding the cultivated area are at the expense of forests, which are under constant pressure from a growing population.

The farming system practised in Upper Indus is basically similar to farming in Nepal, i.e, a combination of terrace cultivation and animal husbandry. But even if the farming system is basically the same, the consequences of practising it in Northern Pakistan are quite different from those described in Nepal. This is because the ecology of Northern Pakistan differs dramatically from that of Nepal. By nature, the upper Indus is semi-arid. Only at elevations that are too high for permanent settlement, about 3000–4000 metres, are natural forests to be found, due to orographic conditions. Thus, the critical factor of settlement is the combined access to water and arable land.

In this region, the effects of population increase on forests are quite contrary to those reported from Nepal. More people need more irrigated farmland, which is supplied by extending existing channels and by constructing new ones. More farmland needs more manure, implying an increase in domesticated animal populations. In turn, more livestock implies increased demand for summer pastures and winter fodder. Summer pastures are plentiful in the high mountains. The bottleneck for increased animal populations is, primarily, the supply of winter fodder. During winter, cattle, sheep, and goats survive on agricultural residues, on browsing the natural semi-arid vegetation, and on foliage gathered from irrigated valley forests. Thus, winter-feeding of an increasing number of domesticated animals is partly solved by extending the area of irrigated forest.

A farming system based on a combination of terrace cultivation and animal husbandry in Northern Pakistan implies a positive correlation between population increase and valley forest cover, according to the formula: "more people – more irrigated forest". This is quite contrary to the "more people – more erosion" formula reported by some from Nepal.

There is no direct link between farming and the high mountain forests in semi-arid Northern Pakistan, since extension of farmland does not imply encroachment into forests. The distantly located mountain forests have been allowed to regenerate with only minor human interference for centuries.

Lately, however, exploitation of the high mountain forests has started. During the last 10 to 15 years, a considerable amount of timber has been taken out. But the extraction of timber is not related to the farming system. The immediate cause is the opening of the Karakorum Highway in the late 1970s and the construction of link roads up the side valleys. Trees, hundreds of years old, are cut and transported to the Karakorum Highway by donkey, jeep, or lately, on high-capacity timber slides, where they are taken over by contractors from the plains. The road has allowed the commercial exploitation of the forests, but the reason for the eagerness with which local people deplete their forests must be sought in their political system.

Formerly, the Indus Himalaya was called Yagistan, "the ungoverned country", referring to the absence of centralized political authority. A mosaic of politically independent valleys and settlements is internally regulated along principles of kinship, ethnicity, and territory. Local disputes and questions of common resource management have been handled by the institution of "jirga": a council of influential men trying to reach consensus on matters in question. However, the role of the jirga is limited to mediation. If consensus is not reached in the jirga, or if the involved parties do not accept the verdict reached, few means of sanction exist. In such cases a solution is left to the relative physical strength of the parties involved.

As yet, the Northern Areas of Pakistan are only partly integrated into the State. Due to the absence of centralized political and judicial authority, every individual family is the architect of its own fortune. The security of a family's assets is in the last instance conditioned by its physical strength, and a family's strength is related to the number of allied partners it is able to mobilise in a conflict. Traditionally, the political process in the area can be viewed as a constant manipulation of social relations in a game of support and opposition.

Recently, however, a new factor has been added to the local power games. When Soviet troops withdrew from Afghanistan in 1988, they left vast numbers of automatic weapons behind in the custody of Afghan warlords. Those weapons, especially the Kalashnikov rifle, became highly valued objects in the Indus Himalaya. A happy owner of a Kalashnikov is equal in strength to five men armed with the old Enfield rifles left by the British.

In a local conflict, a family can mobilize allied partners by investing in social capital through marriage and relationships of reciprocity. Kalashnikov rifles, on the other hand, must be bought for money. Cash is supplied by exploiting the natural resource that became available by the Karakorum Highway – the high mountain forest. Depletion of mountain forests does not worry the tribesmen so much, since it has no direct relevance for their farming practices.

In conclusion, the arms race between autonomous families in Indus Himalaya generates a next to insatiable demand for weapons, reminiscent of the international arms race in the Northern Hemisphere until recently. In the same manner as environmental considerations were ignored during our own postwar arms race, and as they still are in the India – Pakistan war game, Western Himalayan families give priority to their own security and political sustainability, before thinking of a communal ecological sustainability. If it is considered a goal to halt the rate of deforestation in the acephalous regions of Northern Pakistan – and probably also in Afghanistan – solutions must be looked for in the political system.

Hermann Kreutzmann
Institute of Geography, Friedrich Alexander University
Kochstrasse 4, 91054 Erlangen
Germany

Water towers of humankind: approaches and perspectives for research on hydraulic resources in High Asia

The policy document "Mountains of the World: Challenges for the 21st Century" (Mountain Agenda, 1997) refers to mountains as the water towers for the 21st century, and in Mountain Agenda's contribution to Chapter 13 of "Agenda 21", highland waters are defined as a resource of global significance (Bandyopadhyay *et al.,* 1997). This view and estimation holds especially true for the mighty wall of ranges separating the Inner Asian deserts and steppes from the plains of the Indian subcontinent. The Hindu Kush–Karakoram–Himalayan arc provides valuable water resources for some of the most extensive irrigation networks on Earth. The allocation of meltwaters from the nival zones to low-lying deserts has a long-standing tradition and is responsible for the utilisation of arid zones in the vicinity of mountains.

Gravity is the driving force, and rivers are the transportation lines for the valuable resource over long distances. The extension of irrigation networks, construction of dams and generation of hydro-energy have increased over time and contributed to the integration of mountain regions into supra-regional exchange patterns. Among other factors, the interdependence of lowland and highland societies is characterised by the utilisation of water. This resource has become a strategic factor, and conflicts over the rights of users have occurred. Water management is concerned with the energy-efficient transportation of hydrologically exploitable resources from the upper zone to climatically favourable areas where irrigation helps to overcome arid conditions for the cultivation of crops and the watering of meadows.

In other words: human intervention sets the stage for the allocation of water from a wider catchment area to a smaller habitat where the resource is deficient. Emphasis on montane irrigation practices is countered with developments in the lower areas, where a different framework of conditions prevails and development problems are peculiar to those areas. The contrast between highlands and lowlands is highlighted here in order to accentuate the specific formulation of intentions. In dealing with the importance of water from the mountain regions, three factors must be evaluated:

- natural factors and their validity for the environmental framework and technological adaptation processes;
- social factors and their impact on culture, economy and equity;

- institutional factors and their importance for sustainable growth and for the implementation of development projects.

This approach allows an assessment of the contributions of different sectors towards the understanding of water management in mountain communities. Each component of the system requires an elaborate analysis of certain properties linked to hydraulic resources and their utilisation (Kreutzmann, 1996). Singular elements in themselves can only provide insights into limited aspects of the objective. The principal perception gap between the apparent features of a glaciated mountain range and the hidden aspects of water rights (Netting, 1974) and regulations in the associated water-user community aggravates the task for ensuring a holistic approach. Thus linkages and interrelationships among the system components must be investigated.

Projecting such an approach onto certain cases in the High Asian mountain belt, we discover that, on the one hand, we are amidst the water towers of humankind with glaciation over more than a quarter of the surface in some places, such as the Karakoram; and on the other hand, we experience an extreme seasonal scarcity of water availability in relation to crop requirements. The decentralised irrigation systems are embedded in a set of environmental, economic and societal constraints. The existence of locally-invented irrigation systems within the harsh mountain environment has been neglected over long periods, due to the attribution of backwardness and a limited growth orientation toward remote valley societies.

Only within the last decade has awareness grown substantially, and a deficiency of accessible information about their cultural and socio-economic foundations is ubiquitous. In recent years, development agencies implementing integrated rural development programmes have tried to build on local knowledge and to place emphasis on co-operation with farmers in order to serve their perceived needs. In summary, there appears to be an increasing effort to understand the complexity of locally-adapted irrigation systems and the connected societies which could serve as a nucleus for regional development involving a shift from large-scale projects to decentralised activities. One must keep in mind that complexity and variety within the mountain habitat form one of its principal features.

REFERENCES

Bandyopadhyay, J., Rodda, J.C., Kattelmann, R., Kundzewicz, Z.W. and Kraemer, D. (1997) Highland waters – a resource of global significance. In Messerli, B. and Ives, J. (eds.) *Mountains of the World: A Global Priority*. Parthenon, New York, London. pp. 131–155.

Kreutzmann, H. (1996) Wasser als Entwicklungsfaktor in semiariden montanen Siedlungsräumen. Systemansatz und Entwicklungspotential, *Zeitschrift für Wirtschaftsgeographie*, **40** (3): 129–143 [English translation: Water as a factor for development in semiarid mountainous settlement zones, *Applied Geography and Development*, **51**, 1998 (In press)].

Mountain Agenda (1997) Mountains of the World: Challenges for the 21st Century. Mountain Agenda, Bern.

Netting, R. (1974) The system nobody knows: Village irrigation in the Swiss Alps. In Downing, T.E. and Gibson, M. (eds.), *Irrigation's Impact on Society*. Tucson (Anthropological Papers of the University of Arizona 25), pp. 77–82.

Irina Merzliakova[1] and Alexandre Sorokine[2]
[1]Institute of Geography, Russian Academy of Sciences
Staromonetny per. 29, 109017 Moscow, Russia
[2]Regional Science Institute, 4–13 Kita–24 Nishi–2
Kita-ku, Sapporo 001, Japan

Land-use changes and decision making: the Pamiro-Alai case study

This paper presents a selection of results of research conducted within the project "Electronic Atlas of Tajikistan" carried out by the International Mountain Laboratory, Institute of Geography, Russian Academy of Sciences, in 1990–95.

One aim of the study was an analysis of land-use changes in the Pamiro-Alai mountain region during the 1949–1991 Soviet period, as they reflect a general trend toward economic polarisation between the administrative centre and the mountain periphery and a forced shift towards a planned economy, restructuring the resource utilisation scheme and the resulting evolution of local agricultural traditions.

The study area covers the lower and central part of the Surkhob valley in Tajikistan. This area is fragile with regard to exogenic processes due to seismic and anthropogenic activity. The Khait earthquake of 1949 had a magnitude of 10. Settled agriculture was introduced in the region 300 years ago. The population reached 120,000 by the beginning of the 20th century, but has since decreased due to forced resettlement after 1949.

We have created an extensive three-tier database of environmental and socio-economic information with levels that correspond to three different map scales. Basic layers (digital elevation model, morpholithogenic units, exogenic processes, land-use, settlements and road network) were compiled, based on field surveys, aerial photo interpretation and published data. The core of the socio-economic database comprises demographic data and information on the transportation network and crop production. The middle level includes synthetic layers devoted to land suitability, land-use dynamics, and exogenic vulnerability of slopes. Spatio-temporal analysis resulted in a set of maps and cartograms for three specific years: 1949, 1954, and 1991.

Cartograms show ratios between parameters, mean indices and compliance with assessment criteria. They are built on the basis of regional-isation according to groups of watersheds (or districts), these being the traditional basis for community economy. A uniformity of environmental conditions, combination of natural resources, economic specialisations and trends in socio-economic development including modern ones determined the integrity of the regions.

Comparative analysis of the territories by a set of parameters, and the evaluation of changes of their relative importance over time, could be used as an additional illustrative means for decision making. It is intended to identify problems and relative trends during strategic planning of resource utilisation and land-use structure.

Estimated losses of agricultural land due to structural changes in land-use were compared to those inflicted by the earthquake. Some 34% of arable lands not damaged by the earthquake were transformed into poor pastures, wastelands and badlands, with considerable shifts in economic orientation among watersheds, incompatible with the land potential and its modern environmental status.

Analysis of land-use transformation since the beginning of the century brought the conclusion that centralised planning is the most important factor in transformation of the local socio-environmental system. Environmental hazards and resulting depopulation show the cumulative effect that led to the destruction of the intensive traditional agriculture, the development of waste lands and aggravation of the ecological situation.

Land-use was most transformed on the right bank of the Surkhob, which Arandarenko (1878) called "the valley of cereal landscape". Traditionally, all suitable patches of land were ploughed and productivity was high (wheat and barley: nine to ten centner/ha under irrigation, six to seven rainfed). Due to the earthquake and the further transfer of lands to the temporary ownership of southern collective farms, the region lost its agricultural importance and ceased to be the main provider of cereals to the local market in Khait.

Our estimate of the total harvest of cereals (quite stable per year per person) was far greater in 1878 than the World Food Programme (1996) nutrition norms (40 to 73 kg annually per person). Even in 1944 before major restructuring of the economy, collective farms produced about 220 kg of wheat grain per person. In 1991, only two sacks of flour (120 kg) were supplied to each household (average seven persons) by the government (Merzliakova, 1996). This was especially devastating to the regional agriculture when the governmental and collective farms were destroyed. The private sector could hardly compensate for this process on plots of 0.2 to 0.7 ha, in comparison to the average private land ownership of one to two ha in the 19th century.

The currently continuing repatriation of mountain dwellers, the presence of abandoned lands in relatively good condition, and a fall of grain production at national level call for wide-scale reclamation of arable lands. Using the above-described maps and diagrams in planning the agricultural strategy, districts requiring further attention were identified. At a detailed level, the modern status of this potential was estimated in relation to exogenic processes of development and accessibility. Such an analysis for the most promising area within the upper section of Surkhob basin, the Jasman valley, gave an average 15% of abandoned lands that need recultivation before cropping.

A revision of resource utilisation strategies, land-use schemes and decision-making approaches will be required in order to overcome political instability, economic backwardness and environmental degradation. In the event of war, a similar analysis could help to determine the agricultural capacity of the region.

REFERENCES

Arandarenko, G. (1878) Karategin, Military collection, n.5.

Merzliakova, I.A. (1996) Analysis of resource utilisation in mountain region (Pamiro-Alay case study). GIS approach, a dissertation.

World Food Programme (1996) WFP in the CIS, Situation report, n.6, May 1996.

Koko Warner-Merl
International Institute for Applied Systems Analysis (IIASA)
Landstrasse-Hauptstrasse 98/3/16, 1030 Vienna
Austria

The Russian Altai: potential environmental impacts on mountain areas from anthropogenic air pollution

The issue of toxic atmospheric emissions has captured international attention, and efforts towards abatement have been undertaken. However, the environmental impacts and social and political implications of emissions upon mountain regions require further research. Mountain areas have a lower ability to withstand certain levels of anthropogenic emissions without sustaining irreversible chemical change, which can result in a loss of biodiversity or change mountain ecosystems in ways that render both natural and economic processes less viable. In addition, mountain areas such as the Altai are also coming under development pressure as the need for natural resources, recreation areas, and industry increase. The Altai is a sparsely populated, resource-rich area with enormous development potential. Yet the environmental impacts of such development (and air pollution by way of extension) could be negative and irreversible, with long-term consequences for soil, the unique ecosystems of this complex area, and the entire region which depends upon the hydrological systems rising in the mountains.

Hard-hit by Russia's painful economic transition which closed many industrial plants, the Altai receives large subsidies and is targeted for greater economic development (Bradshaw and Shaw, 1996). While large parts of Siberia are being deserted, the Gorno Altai experiences inward migration of 1.1 to 2.8% annually (Bradshaw and Palacin, 1996). The area underwent a 4% increase in population between 1989 and 1995, and has both some of the highest birth and infant mortality rates and lower than average life expectancies in Siberia: 55.5 years for men and 68 for women in the Gorno Altai in 1994 (Goskomstat, 1994a). The Altai Kray and Gorno Altai are relatively less-industrialized Siberian areas. Rural and agricultural output plays a significant role in the economy of the Gorno Altai, from which 1,202,000 rubles are generated per capita, as opposed to just 57,000 rubles for industrial activities (Goskomstat, 1994b).

The Russian Federation has gathered data on emissions in order to assess the state of the Siberian environment. An initial attempt to calculate critical loads in the region shows peculiar patterns of exceedances in Siberia's southern-central mountains, particularly the Altai and Sayan (Bashkin *et al.*, 1995). While human activities there do not produce high quantities of toxic emissions, surrounding areas – particularly the industrial centres in Kemerovo and Novosibirsk – contribute significant atmospheric pollution. With steeper

155

gradients, different types of soils, less vegetation, and more severe weather, the Altai range absorbs, and is disproportionately affected by, these industrial emissions. The central issue concerning air pollution for the Altai and numerous other mountain regions lies in protecting these sensitive environments from such intangible external destructive forces. Reducing air pollution in industrial centres in Russia would have a beneficial effect on the air quality and environmental health of the Altai and other Siberian mountains.

These conclusions were reached by using a pollution database made available by the Russian Federation to analyze the reported environmental impacts of emissions in the region. This gives an overview of the potential risks of those emissions to the Altai. Critical loads and their exceedances are analyzed for the Russian Altai. The presentation of unpublished and hard-to-come-by information regarding Russian industrial emissions in the Altai region constitutes a key part of this study, listing over 100 organic and inorganic chemical substances. While these threshold value calculations are somewhat rudimentary and on a large scale of 10,000 km^2 as opposed to under 150 km^2 in European models, (Posch *et al.*, 1997) ongoing research in air pollution in Siberia will yield better results. Particular necessities are for "protection isolines", critical load thresholds for combinations of chemicals, and better data quality for the abatement of harmful chemical emissions within the political and social climate.

This case study brings several themes to the foreground, which other mountain areas share. First, mountains suffer from environmental degradation whose sources often lie outside the mountains. Addressing the sources of pollution in industrial centres must be the next step in protecting mountain areas. Second, although industrial emissions affecting the Altai mountains have subsided in recent years, along with industrial decline, regional development policy will play a key role in the protection or destruction of sensitive mountain ecosystems. Development strategies must be modified for their host environments and, in mountain regions where air pollution has particularly noxious effects, deliberate shifts away from fossil fuels, as industrial energy sources could dramatically improve air quality. In addition, where financially possible, better filter technologies could be installed in old plants to reduce air emissions. Finally, it will become increasingly necessary to make mountains an international priority, to reduce pollution levels elsewhere, and to recognize the potential downhill and downstream risks incurred by damaging mountain ecosystems through air pollution.

REFERENCES

Bashkin, V.N., Kozlov, M.Ya., Priputina, I.V., Abramiychev, A.Yu. and Dedkova, I.S. (1995) Calculation and mapping of critical loads of S, N and acidity on ecosystems of Northern Asia. Institute of Soil Science and Photosynthesis, Russian Academy of Sciences, Moscow.

Bradshaw, M.J. and Palacin, J.A. (1996) An atlas of the economic performance of Russia's regions. Russian Regional Research Group, School of Geography and Centre for Russian and East European Studies, Working Paper 2. University of Birmingham.

Bradshaw, M.J. and Shaw, D.J.B. (eds.) (1996) Regional problems during economic transition in Russia: Case studies. Russian Regional Research Group, School of Geography and Centre for Russian and East European Studies, Working Paper 1. University of Birmingham.

Goskomstat (1994a) The Demographic Yearbook of the Russian Federation, Goskomstat, Moskow.

Goskomstat (1994b) Sravitel'nye pokazateli ekonomischeskogo polozheniya regionov Rossiiskoi Federatsii, Goskomstat, Moscow.

Posch, M., Hettelingh, J-P., de Smet, P.A.M. and Downing, R.J. (eds.) (1997) Calculation and mapping of critical thresholds in Europe: Status Report, 1997. Coordination Centre for Effects, National Institute of Public Health and the Environment, RIVM Report No. 259101007. Bilthoven, Netherlands.

Zheng Du
Institute of Geography, Chinese Academy of Sciences
P.O. Box 9719, Beijing 100101
People's Republic of China

Studies of the environments and development of the alpine scrub and meadow zone on the Qinghai-Xizang (Tibet) Plateau

Located in the middle-eastern part of Qinghai-Xizang (Tibet) Plateau, the natural zone of alpine scrub and meadow is a transitional area from the deep gorges to the plateau proper. This zone, with a mean temperature of 6 to10 °C in the warmest month, and annual precipitation of 400–800 mm, belongs to a plateau subpolar humid/subhumid climate. Main types of alpine meadow include *Kobresia* meadow, herbaceous meadow and swamp meadow. In relation to physical environments and natural ecosystems, no corresponding zone exists in lowland regions anywhere else on Earth (Zheng, 1996).

Regarding human activities, this differs clearly from other natural zones of the sub-polar belt on the plateau. It is roughly estimated to have an area of 269,000 km^2: 0.7% of the total area of the plateau. The population density of 3.0 persons/km^2 is much higher than those of the alpine-steppe and alpine-meadow-steppe zones of the plateau with negligible impact of human activities (Zheng, 1996).

The natural zone of alpine scrub and meadow is important for animal husbandry, being particularly favourable for grazing yaks and sheep of Tibetans. The grassland resources are characterised by alpine meadows with a short growth of herbage, high proportion of edible forage grass, low production, high value of nutrients, and high utilisation ratio. Taking the Golog Autonomous Prefecture of Qinghai as an example, the total area of alpine meadow grasslands amounts to 5,743,700 ha – 89% of the total area of grasslands in the Prefecture – while the available area of grasslands amounts to 5,348,700 ha.

Concerning the natural grasslands, the main limiting factors for the development of animal husbandry are an unbalanced distribution of grassland in time and space, with a short growing season and long period of withered grass, overloading of pastures in winter-spring (seven to eight months) and their inadequate utilisation in summer-autumn (four to five months). To maintain the stable development of animal husbandry, grasslands should be demarcated into distinct pastures and used in turn. A prolonged grazing period on high pastures is encouraged in the summer-autumn season, because it helps reduce the pressure on winter-spring pastures. The grazing intensity of pastures should be maintained at a level of 45%–55%, in order to promote the regenerative ability of herbage (Zhou Li *et al.* 1995).

The total number of livestock of Golog has increased from 1,467 million head in 1953 to 2,503 million head in 1989. Overgrazing and trampling bring about serious degradation of grassland, with a decrease of edible herbage and an increase of detrimental and poisonous weeds, thus reducing productivity. There are degraded grasslands of 2.86 million ha – 42.7% of the total area of grasslands in Golog – and degradation by overgrazing has been intensified by recent reforms which have placed livestock in private hands, whereas the grasslands are held in common. The solution is to expand the double-responsibility system by which grazing areas and animals are assigned to the same households, and to create a self-interest on the part of producers to maintain the balance between land and livestock (National Research Council, 1992).

Pika, rodents and caterpillars have severely damaged the meadow vegetation, causing grassland degradation and the formation of 'Heitutan' (literally, black earth sands). The grassland destroyed by rodents and pikas amounts to 1.44 million ha: 21.5% of the total grassland area in Golog. Snow accumulation in winter and spring also has a serious effect on animal husbandry in the zone.

Restoration measures are needed for the management of degraded grasslands. Experiments show that by restoring the grasslands for two to three years, the population, density and coverage of forage grass are increased, thus changing the structure of plant communities, improving the environment, enhancing the biomass and raising the productivity of pastures (Zhou *et al.*, 1995).

The population structure of livestock should be adjusted, to enhance the utilisation of grasslands and the marketing of livestock. The development of pastoral animal husbandry should be promoted through maintaining enough breeding female livestock and reducing the pressure on winter pastures. Consistent with a sustainable maximum grazing intensity of the two-season pastures, an optimum grazing system should be established, aimed at the maximum productivity of livestock and economic profit. (Zhou Xingmin, 1996).

In addition, several key measures should be taken in pastoral areas, such as the improvement of the breeds of domestic livestock, the enhancement of investment for animal husbandry, education for the training of herdsmen, adjustment of the productive structure, establishment of a fodder and livestock products processing factory, and strengthening of capacity building.

REFERENCES

National Research Council (1992) *Grassland and Grassland Sciences in Northern China*, National Academy Press, Washington DC.

Zheng Du (1996) A preliminary study on the zone of alpine scrub and meadow of Qinghai-Xizang (Tibetan) Plateau, *The Journal of Chinese Geography*, **6** (3): 28–38.

Zhou Li *et al.* (1995) Studies on optimum stocking intensity in pasturelands of alpine meadow (1–4), In *Alpine Meadow Ecosystem Fasc. 4*, Science Press, Beijing, pp. 365–418 (in Chinese).

Zhou Xingmin (1996) Rational use of grassland resources and sustainable development of animal husbandry in Qinghai, In He Xiwu, *et.al.* (eds.), *Proceedings of Symposium* on Resources, Environment and Development of Qinghai Province, China Meteorological Press, Beijing, pp. 110–116 (in Chinese).

Zhou Xingmin *et al.* (1995) Present condition of degenerated grassland, regulation tactics and sustainable development in Qinghai-Xizang Plateau, In *Alpine Meadow Ecosystem Fasc. 4*, Science Press, Beijing, pp. 263–268 (in Chinese).

Zhu Li-ping
Institute of Geography, Chinese Academy of Sciences
P.O. Box 9719, Beijing 100101
People's Republic of China

The main natural hazards on the Qinghai-Tibet Plateau

Introduction

The Qinghai-Tibet Plateau, with an area of 2.5 million km^2 and an average elevation over 4,000 m, is the most concentrated region of high mountains on Earth. As it is characterized by complex geological and geographical conditions, different climatic types and a fragile and harsh ecological environment, different natural hazards occur unceasingly on this Plateau and its peripheral areas (Zheng *et al.*, 1996). The costs of damage from natural hazards are far lower than in the economically developed regions, due to the backward social economy and sparse population on the Plateau. However, with the gradual development of the economy, human activities continue to increase in this fragile environment. This causes natural hazards to occur frequently, while the wealth of the society increases (Wu, 1996). In some cases, the loss and damage are difficult to restore.

Distribution features of the main natural hazards and their damage

There are many natural hazards on the Plateau, including landslides; debris-flows; frost and snow; glacial floods; river valley desertification; grassland degradation by pika and zokor; and earthquakes. Based on the geological and natural geographical features and the socio-economic development in different areas of the Plateau, major hazards may be distinguished according to their distribution and the degree of damage.

- Landslides and debris-flows occur mainly on the south slopes of the Himalaya and the Hengduan Mountains. These areas are situated in the peripheral areas of the Plateau, where a great number of valleys have been cut by rivers. The steep slopes and rapid currents are very favourable to the formation and development of landslides and debris-flows. As it is the main forest area of the Qinghai-Tibet Plateau, and has considerable water energy resources, this area has huge economic development prospects (Chen *et al.*, 1992). However, the landslides and debris-flows seriously imperil roads and other engineering works, and limit economic development.

- Snow and frost hazards occur mainly on the broad plateau surface. As these are the main natural grazing grounds, winter snow is a frequent hazard in this area, and directly destroys the productive base of the local people (Lin,

1992). Under the action of the frost-thaw cycle in the permafrost area, roads and other engineering works are often destroyed (Zhang and Chen, 1992).

- Pika and zokor hazards in grazing areas affect 17 million ha of the 120 million ha of the usable natural grazing areas on the Plateau. This is mainly due to the increased human activities which lead to over-grazing and then cause the degeneration of the pasture land and the subsequent large area of pika and zokor damage (Fan, 1992).

- Earthquake hazards occur frequently on the Plateau and its peripheral areas. This is because of very active neo-tectonic movement, as seen in the uplift of the Plateau since the late Cenozoic. However, due to the low population and backward economy of the Plateau, the frequent and large earthquakes which occur there result in lower losses than in the developed area (Li *et al.*, 1996).

Study and management of the main hazards on the plateau

With socio-economic development on the Plateau, there have been many studies of the mechanisms of hazards and their control. In the southern and south-eastern parts of the Plateau, debris-flows have been studied both in theory and through large-scale fieldwork. The frequency and scale of landslides and debris-flows in some key areas have been effectively controlled by soil-water management, bans on unorganised felling, and an increase in engineering approaches. However, due to the extremely steep valley landforms and heavy precipitation, landslides and debris-flows occur everywhere. These kinds of hazards are still the main threats to economic development, especially for safety of the roads.

Snow hazards destroy the grazing ground in winter and lead to the death of many livestock due to the shortage of natural herbage. Thus, the accurate prediction of snow is very important. At present, the range of snow hazards can be predicted using remote sensing and geographic information system (GIS) methods, but not the depth of snowcover. Frost continues to destroy roads in the permafrost areas on the Plateau. In the past, some 10 million RMB (1.2 million US dollars) have been spent yearly in repairing the Qinghai-Tibet Highway. With the study of the freeze-thaw mechanism, and progress with engineering experiments, this hazard is being managed by engineering techniques.

Pika and zokor hazards are difficult to control as they occur over such large areas and have been influenced by policies of local governments. To date, such hazards have been controlled only in some demonstration areas, based on the strategy of encouraging the people to "appreciate pasture", and some chemical methods. More effort is needed for the extension of effective management methods.

Earthquakes are frequent, but their damage is small due to the undeveloped economy and low population. The study of earthquakes and their activity has contributed to geological theories, while also providing a good basis for future development and construction on the Plateau.

REFERENCES

Chen Zisheng, Wang Chenghua and Kong Jingming (1992) Landslide disasters and macro-prevention methods in China. In: Shi Yafeng, Huang Dingcheng and Chen

Banqin (eds.), *Disaster Conditions Analyses and Hazard Reduction Counter-measures for Natural Hazards in China*. Wuhan: Hubei Science & Technology Press, pp. 307–313. (in Chinese)

Fan Naichang (1992) The tendency of major populations of injurious mice and their control in Qinghai-Tibet Plateau. In: Shi Yafeng, Huang Dingcheng and Chen Banqin (eds.), *Disaster Conditions Analyses and Hazard Reduction Counter-measures for Natural Hazards in China*. Wuhan: Hubei Science & Technology Press, pp. 416–420. (in Chinese)

Li Bingyuan, Li Juzhang and Wang Jianjun (1996) Areal association of natural hazards in China. *Acta Geographica Sinica*, **51**, (1), 1–11. (in Chinese with English abstract)

Lin Zhenyao (1992) Climatological analysis of snowstorm damage in Xizang. In: *China Society for Qinghai-Tibet Plateau* (ed.), Proceedings of the First Symposium on the Qinghai-Xizang Plateau, Beijing: China Science Press, pp. 228–234. (in Chinese)

Wu Yongsen (1996) Environmental pressure of the Plateau development. In: Luosang Lingzhiduojie (ed.), *Introduction to the Environment and Development of the Qinghai-Tibet Plateau*. Beijing: China Tibetology Press, pp. 137–153. (in Chinese)

Zhang Xiangsong and Chen Xiaobai (1992) Studies on disasters of ice-snow and engineering freezing and their prevention. In: Shi Yafeng, Huang Dingcheng and Chen Banqin (eds.), *Disaster Conditions Analyses and Hazard Reduction Counter-measures for Natural Hazards in China*. Wuhan: Hubei Science & Technology Press, pp. 253–258. (in Chinese)

Zheng Du, Yang Qingye and Liu Yanhua (1996) Natural environment and zones differentiation. In: Sun Honglie (ed.), *Formation and Evolution of Qinghai-Xizang Plateau*. Shanghai: Shanghai Science & Technology Press, pp. 262–323. (in Chinese)

Rita Gardner
Royal Geographical Society with the Institute of British Geographers
1 Kensington Gore, London SW7 2AR
UK

Introduction: Eastern and Southern Asian mountains

Arguably the most central and pressing issues in the mountain regions of monsoonal Eastern and Southern Asia concern natural resources management. Such issues have been high on research agendas since the mid 1980s. Over time, the initial simplistic assumptions about population pressures interacting with the intensely active environmental systems which characterise this region are slowly being modified. The papers in this section illustrate this process of modification of paradigms, such that greater awareness is given to indigenous knowledge and the efficacy of traditionally-developed systems of resource management. There is greater awareness that human impacts have not necessarily been as great or as damaging as previously hypothesised. We see increasing numbers of documented examples of positive local adaptation to change, and increasing awareness of the diversity and complexity of both systems and their responses to change. Perhaps the overriding message is that physical, economic, social, and political systems are inextricably linked when dealing with the central issues of sustainable development, i.e., of natural resource management, whether of soil, water, or vegetation resources. It is the reinforcement of these new paradigms which comes across most strongly as a theme to the papers in this section. In practical terms, this underlines the need for local involvement in, and responsibility for, sustainable development.

Nevertheless it is always worth remembering that, given the scale of these landscapes and the energy of the system, there is a much wider context to the debates surrounding the impact of human activity upon ecological and landscape systems. The magnitude and frequency of natural events and their translation into natural hazards affecting the livelihoods, and at times the very life, of the local populations is an important issue. It is one addressed by Fort, who argues that hazards in Nepal are largely the result of natural factors. She draws particular attention to their contribution to removal of materials from mountain systems and to their influence, in both the short and long term, on creating zones of high susceptibility to instability. This scale of instability and geomorphic activity is seen as providing a context to understanding more localised responses to human actions in the environment.

Natural processes in mountain systems are also considered by Lasco and Pulhin in their assessment of land-use change and its implications for the actual and potential role of forests in mitigating climate change through carbon sequestration. In a world faced with global warming and increasing CO_2 emissions, far greater understanding and quantification of the carbon cycle is still required. The analysis of carbon storage and loss from different vegetation communities within the mountains shows the importance of agroforestry farms in the sequestration of carbon, in contrast to the old growth areas which are the predominant carbon store.

A more traditional consideration of the management of forest resources is given by Seeland, who discusses the national park management regime in Bhutan. A comprehensive historical background illustrates how policy has developed to protect areas of forest and biodiversity resources. However, the case study also illustrates the conflicts that can arise between local communities and central governments following demarcation, and describes how these are being overcome through informal local collaboration and participation.

As stated above, much recent research has started to question implicit assumptions about speed of change, and simplistic hypotheses of cause and effect. It is increasingly showing the depth of indigenous knowledge and the nature of indigenous response to changing natural resource availability. In the case of Nepal and Ladakh, Smadja shows the importance of using historical data to question common perceptions, and in particular those concerning recent rapid deforestation owing to human pressure on the land. This work also demonstrates the adaptive strategies employed by farmers in terms of fodder provision and reducing pressure on limited forest resources.

Water resources are equally as important as forest resources within a Himalayan setting, and Kreutzmann makes the point that the Himalayan region provides water resources for extensive irrigation systems and a significant proportion of the world's population. He develops an understanding of water management in the mountain communities, contrasting the potential total availability of water resources with the spatial differences in distribution, and the way in which water supply is ultimately controlled at a local level by a combination of physical, economic and societal factors. He highlights the importance of locally developed indigenous systems for water management and the need to understand the complexities of these systems.

A similar theme emerges in a different setting in Brown's paper on pasture management in Nepal. Traditionally these systems have been well managed by local indigenous institutions, however a combination of demographic, political and economic changes have placed substantial stress on these. The result is a complex pattern of differential response to such stresses. Rapid change is also evident in the Daba mountains of China and the Ifugao mountains of the Philippines described by Saint-Pierre. In both of these regions, different groups of stakeholders agree on the need to maintain terraces. However, there are conflicts over the uses of the terraces for various types of agriculture, agroforestry, or forestry. Local and national institutions play diverse roles in mediating in these disputes.

The paper on the Republic of Korea by Kim and Youn serves to illustrate the differences between mountain systems. The historical trends of land-use change are examined as a background to future sustainable development. This paper illustrates some of the issues arising from substantial and rapid depopulation of mountain regions, an effect only just beginning to be felt in parts of the Nepalese and Indian Himalaya. A broad consensus on the utilisation of forest resources is seen as essential to sustainable development in this region.

The paper by Ramakrishnan helps to draw together the many themes explored in the other papers. The nature of the close connections in the traditional upland societies of Asia between forest ecosystems and local populations, with forest use and management being integrated into the farming systems, is used to illustrate the fact that ecological issues cannot be divorced from social, economic and cultural ones. Arguing that humans are traditionally well integrated into the ecosystem function, Ramakrishnan points out that this contrasts with the common perception of exploitation. However, increasing population pressure is seen to be causing greater levels of perturbation in the ecosystems, leading to the formation of complex agroecosystem types. He argues the case for a diversified landscape with a combination of quasi-natural ecosystems existing alongside agroecosystems as a way to maintain bio-diversity. The over-riding message is that sustainable development is inevitably involved with local participation and with the integration of ecology and social science.

Taylor Brown
Faculty of Social and Political Sciences, University of Cambridge
563 Kings College, Cambridge
United Kingdom

The persistence of the commons in Northeast Nepal

Two questions underlie this paper. First, what are some of the major threats to indigenous pasture management institutions in mountain regions? Second, why have some indigenous institutions been more resistant to collapse than others? To address these questions, this paper draws upon fieldwork conducted in the high Himalayan valleys surrounding Mount Kangchenjunga in Northeast Nepal.

Like many agro-pastoralists, the ethnic Tibetans living in the shadow of Kangchenjunga have historically managed pasture resources as common property and have developed indigenous institutions to regulate access to and distribution of pasture products. Scarce and unreliable winter grazing and fodder supplies form a bottleneck in the local agro-pastoral production system. Indigenous institutional arrangements in Kangchenjunga effectively stretch this bottleneck by preventing overgrazing of local pastures, and protecting and distributing fodder resources. In addition, these institutional arrangements also help to minimise intra-household competition and conflict over resources. Specifically, indigenous institutions provide for rotating and resting pastures, closing areas near agricultural lands to grazing during the growing season, and limiting the fodder harvest to prescribed days. Collective decisions governing these institutions are made through meetings of all resource users. Cattle bans and fodder cutting bans are monitored and enforced by watchmen, and conflicts over resources are resolved through a range of mediation and arbitration practices.

Over the past several decades, the Kangchenjunga region has faced unprecedented demographic, political, and economic changes. These changes have placed both direct and indirect pressures on indigenous institutions. First, endogenous population growth, combined with an influx of Tibetan refugees and other immigrants, has nearly doubled the number of households in most settlements in Kangchenjunga.

Second, new markets for tourism, trans-Himalayan trade, carpet weaving, and dairy products have dramatically altered what was until recently a subsistence agro-pastoral economy. As a result, local people have become increasingly integrated into regional and international cash economies, and local resources are used to produce for markets far beyond Kangchenjunga.

Third, although residents of the region have historically exercised a great deal of political autonomy, state policies and agents have recently extended their reach to the local level. Elected village committees have replaced hereditary headmen, and police posts, courts, and other government offices have been established at the district and local level. Attempts at homogenisation of land tenure and taxation throughout the kingdom have brought Kangchenjunga into a closer administrative relationship with the central government in Kathmandu. In addition, since the early 1990s, factionalism along political party lines has increasingly divided some communities.

In some settlements in Kangchenjunga, these economic, political, and demographic changes have undermined indigenous resource management institutions. In contrast, in other settlements indigenous institutions persist, and continue to manage common pasture and fodder resources despite these pressures. For instance, in the settlement of Yamphudin, institutions which once regulated access to, and distribution of, pasture resources have broken down, and wealthy and powerful villagers have privatised the commons. On the other hand, in the nearby settlement of Ghunsa, indigenous institutions continue to function, and pastures remain common property.

For both communities, a pivotal moment for local institutional arrangements was the first government land tenure survey of the region in the late 1980s. The residents of Yamphudin and Ghunsa responded to this land survey in dramatically different ways. In Yamphudin, the land tenure survey was the death knell for common pastures and local resource management institutions. In Ghunsa, the same land tenure survey did not trigger the privatisation of the commons and hasten the collapse of indigenous institutions.

Although there is a range of factors which could account for these different outcomes, we focus on two major issues: equality of access to new economic opportunities, and the continued effectiveness and legitimacy of local governance arrangements.

Over the past several decades in Yamphudin, differential access among households to new markets for dairy products and tourism has led to increasing economic inequality among resource users. Although increasing market integration has fundamentally altered household livelihood strategies in Ghunsa, market integration has not yet created the profound asymmetries of wealth and power now found in Yamphudin.

Political pressures have also affected Ghunsa and Yamphudin differently. Local leadership, decision making, and dispute resolution mechanisms in Yamphudin have failed to adapt to the changing political climate. In Ghunsa, in contrast to Yamphudin, there has been continuity of local political leadership, collective decision making arrangements, and dispute resolution mechanisms. Residents have been able to creatively resist and re-cast external interventions.

In Yamphudin, institutions governing the commons were fatally weakened by increased asymmetries of wealth and power and the breakdown of local leadership, collective choice, and dispute-resolution mechanisms. Thus, when government surveyors passed through, it proved relatively easy for the wealthy and powerful to over-ride the objections of others, and to privatise the commons.

But in Ghunsa, wealth and power remained relatively symmetrical despite market integration, and customary leadership and decision making arrangements continued to oversee the indigenous pasture management regime. There was relatively little incentive for villagers to buy titles from the government surveyors, and those who were tempted to do so faced significant social pressure to change their minds.

Monique Fort
Université Paris 7-Denis Diderot, Case 7001
GHSS, 2 Place Jussieu, 75251 Paris
France

The role of large magnitude events in the distribution of natural hazards in the Nepalese Himalaya

High mountains are typically characterised by extreme topographic and bioclimatic gradients. Their intense geomorphic activity, with higher uplift and denudation rates, reflects the combination of continuing tectonic building and efficient morphoclimatic processes, including glacial action. Natural hazards (earthquakes and/or volcanism, mass-wasting, floods etc.) occur at all temporal and spatial scales. Their damaging impact on growing populations makes their prediction an urgent need for affected societies. Examples in the Nepalese Himalaya show that the distribution of present hazards is very often associated with evidence of pre-existing, larger magnitude, lower frequency, geologically-recorded, hazardous events, thus addressing the nature of preventive decisions to be taken.

The Himalaya, a still growing mountain range created by the collision between the Indian and Asian plates, is eroding at a very fast rate, as attested by the volume of sediments annually trapped in the large sedimentary reservoirs of the Indus and Ganga-Brahmaputra fluvial, deltaic and submarine systems. This erosion rate (2500 t/km^2/yr) is mainly the result of rapid uplift (4–8 mm/yr) and monsoon climate, both forcing a constant adjustment of landforms by river flooding and dissection, and by mass-wasting. These are particularly well developed where uplift rates are the highest, for example in the Mahabharat Range and the Higher Himalaya, and where the bedrock lithology is the weakest, such as weak schistose rock units and disintegrated granitic bodies of the Lesser Himalaya (Fort, 1987a; Gerrard, 1994), or where the bedrock structures and cleavages are steeper (>30°N dipping micaschistic and gneissic slabs at the lower and upper parts of the Himalyan crystallines). Slope amplitude and shape, and impluvium size, are additional factors influencing the present distribution of natural hazards.

Another understated, yet prevailing factor is the legacy of past geomorphic evolution, expressed as regolith or paleolandslides and/or floods, the significance of both being quite different. On one hand, the fact that present landslides may affect slope regolith indicates a certain degree of former stability which is presently unbalanced by change in the geomorphic system, induced by either climatic or anthropogenic causes. On the other hand, pre-existing, larger magnitude hazardous events, some of which occurred several tens of thousand years ago, can still have strong impact on the present geomorphic systems. Old, gigantic landslides of several 10^6 km^3 in volume, recorded in various places in

Nepal Higher Himalaya, are thus characterised by zones of shattered and deeply disturbed material (Weidinger and Schramm, 1995; Fort, 1998) where the recurrent landsliding and gullying and the intensity of current geomorphic processes differ strikingly from those of adjacent, unaffected zones. Similarly, former catastrophic floods (as hyperconcentrated stream flow and/or as debris flows) that occurred in Nepal some tens, hundreds or thousands of years ago have left not only large, rather persistent boulders in the landscape (Yamanaka and Iwata, 1982; Fort, 1987b), but also a huge mass of loose, smaller size material that is temporarily stored and can be easily removed by further, lower magnitude events, making these latter more damaging hazards. The causes of these floods are various, either induced by glacial lake (Mool, 1995) or landslide-dammed lake outbursts (Higher Himalaya) or by exceptional climatic events such as rainfalls of very high intensities and duration (Fort, 1997) that are more characteristic of the Lesser Himalaya.

It is clear that present hazards in the Nepalese Himalaya are strongly controlled by natural factors. Some are persistent in the long term (regional uplift/dissection rates, bedrock nature and structure, etc.) and thus are more predictable than others whose occurrence cannot be anticipated (seismicity, precipitation distribution, rate of glacier melting, etc.). The occurrence of smaller size, more frequent, short-term hazards is controlled at local scale by other factors: either natural and rarely predictable, such as the amount and intensity of precipitation, underground water flows etc. ; and/or human factors associated with population density, land-use change, road construction, etc. (Brunsden *et al.*, 1981) which create additional potential and more predictable triggers for hazards.

Among various parameters, the distribution of high-magnitude, naturally-induced, events appears to provide the best key to the present understanding of natural hazards. Not only do these events contribute to a large transfer of materials across and out of the mountains, they also indirectly control the present zoning of geomorphic instabilities, the nature and intensity of current processes, and thus the distribution of potential hazards. Studies on hazards should therefore include estimates of sediment yields and geomorphic remodelling of the landscape by past and present extreme events, so that their impact can be predicted, and therefore be accurately considered in land-use management and project implementation in these developing countries.

REFERENCES

Brunsden, D., Jones, D.K.C., Martin, R.P. and Doornkamp, J.C. (1981) The geomorphological character of part of the Low Himalaya of eastern Nepal. *Zeitschrift für Geomorphologie*, **37**: 25–72.

Fort, M. (1987a) Geomorphic and hazards mapping in the dry, continental Himalaya: 1;50,000 maps of Mustang district, Nepal. *Mountain Research and Development*, **7** (3): 222–238.

Fort, M. (1987b) Sporadic morphogenesis in a continental subduction setting: an example from the Annapurna Range, Nepal Himalaya. *Zeitschrift für Geomorphologie, suppl.-Bd* **63**: 9–36.

Fort, M. (1997) Développment et aléas naturels dans l'Himalaya du Népal: la crue de juillet 1993, bassin versant de la Bagmati. *Géomorphologie*, **1**: 43–58.

Fort, M. (1998) Large scale geomorphic events in the Nepal Himalaya and their role in the evolution of the landscape, Fourth International Conference on Geomorphology, Bologne (Italy), Abstract volume: 169–170

Gerrard, J. (1994) The landslide hazard in the Himalayas: geological control and human action. *Geomorphology*, **10**: 221–230.

Mool, P.K. (1995) Glacier lake outburst floods in Nepal. *Journal of Nepal Geological Society*, **11** (special issue): 273–280.

Yamanaka, H. and Iwata, S. (1982) River terraces along the middle Kali Gandaki and Marsyangdi Khola, central Nepal. *Journal of Nepal Geological Society*, (special issue), **2**:95–112.

Weidinger, J.T. and Schramm, J.M. (1995) A short note on the Tsergo Ri landslide, Langtang Himal, Nepal. *Journal of Nepal Geological Society*, **11** (special issue): 281–287.

Eui-Gyeong Kim[1] and Yeo-Chang Youn[2]
[1]Department of Forestry, Gyeongsang National University
Chinju 660–660701, Korea
[2]Department of Forest Resources, Seoul National University
Suwon 441–744, Korea

Historic and current issues of mountain development in the Republic of Korea

This paper aims to present an overview of historical trends of land-use in mountain areas in the Republic of Korea, and to identify the current policy problems which must be settled for the sustainable development of mountain regions. Historical trends in mountain development and its problems were analysed using published statistical and research data together with field surveys.

One mountain development issue in Korea is the conversion of forests to other land-uses, thus continuously decreasing forest land. Forests were converted to cropland to improve food self-sufficiency by the early 1970s, and between the mid-1970s and mid-1980s a large area of forest land was converted to grassland for pasture. From the later 1980s, the conversion of forest to non-agricultural uses increased rapidly, for resort development such as golf courses and skiing areas.

These shifts to other land-uses have many problems. The conversion of forest land to agricultural use failed because of excessive reclamation costs and low profitability. The conversions to non-agricultural land-use for resort construction have also revealed many problems, including watershed contamination and environmental destruction. Ecological and physical instability have resulted from construction processes and excessive use of chemical fertiliser.

In the 1990s, the Korean government is considering mountain development for other land-uses, such as urban land extension, as a measure to prevent the conversion of agricultural land. The government judges that the conversion of agricultural land is bad in the light of low food self-sufficiency, and is therefore promoting a policy to convert forest land to urban uses. Considering the lack of success in mountain development from environmental and economic standpoints, it is to be expected that the establishment of a model for sustainable development will become a new issue in mountain development and forestry in Korea.

A second issue relates to the management of forest resources. The strong afforestation policy resulted in the complete replanting of denuded areas. The newly planted forest area, under monoculture, covers 32 percent of forest land. This forest is in need of effective silviculture, but this has been abandoned because of low profitability because of very cheap imported timber. If the structure of the timber market remains unchanged, most mountain forests in

Korea may become unprofitable for timber production, and will be abandoned in the absence of management.

A final issue relates to societal change in mountain areas. The depopulation of these areas is very serious, with populations decreasing 46.7% during the past 30 years. This rapid decrease has brought various problems in the management of domestic forest resources, such as the shortage of labour force. In addition to depopulation, the increasing proportion of elderly people makes the labour supply for the utilisation of natural resources much worse.

Some of the important needs for guaranteeing sustainable mountain development concern the development methods of domestic resources, and societal conflicts. In order to survey options for domestic resources development, 189 households in Doam-myeon were interviewed in 1993. Face-to-face interviews were used to determine residents' opinions on methods for developing such resources, and their reactions to governmental policies for achieving the conservation of the area's unique ecosystem. The results show that regional development through external investment is not much favoured. However, regional development in Korea has mainly been implemented through outside investors, and the research showed that the indigenous rights of local residents have been damaged with little benefit to their income.

Another field survey on societal conflicts was undertaken by interviewing 201 persons who visited Mt. Baek-un in 1994. This survey, on the appropriateness of the government's designation of a Natural Ecosystem Conservation Reserve on Mt. Baek-un, revealed differences of opinion between the general public and the local residents.

In conclusion, in order to guarantee sustainable mountain development in Korea, the formation of a national consensus on the utilisation of forest resources is most important, after which detailed political options for its realisation must be sought. Forest resources management, local development methods, the government's support policy for continuous forest investment, and criteria for the conversion of forest lands to other land-uses are some of the options to be investigated.

Rodel D. Lasco[1] and Florencia B. Pulhin[2]
[1]Associate Professor and Director, Environmental Forestry Program
UPLB College of Forestry, College, 4031 Laguna, Philippines.
[2]University Researcher, Forestry Development Center
UPLB College of Forestry, College, 4031 Laguna, Philippines.

The mountain ecosystems of the Philippines: opportunities for mitigating climate change

The paper analyses land-use change in the Philippines during three periods: the past (1500s to 1990s), the present, and the future up to 2015. From the 1500s to the present, the Philippines lost about 21.3 M(million) ha of tropical forests. This is equivalent to 3.7×10^9 tons of carbon (C) released to the atmosphere. More than 70% of this (2.7×10^9 tons) was released in this century alone.

At present, 15.88 Mha of the Philippines could be classified as mountain areas ($\geq$18% slope) (Forest Management Bureau, 1996). The six major land-uses are: old-growth/protected forests, second-growth forests, brushlands, grass-lands, tree plantations and agroforestry farms.

In general, 2.70 Mha of old-growth forests contain 307 M tons of C and absorb 4.1 M tons annually (Lasco, 1997). The main value of these forests lies in the huge amount of carbon stored in the biomass. However, although legally protected, threats of further destruction are ever-present because of limited resources.

Second-growth production forests (3.40 Mha) harbour 298 million tons of C and capture 0.4 M tons net of C yearly. Harvesting and deforestation are the main pathways of carbon loss, with 0.3 and 8.8 M tons/yr respectively.

Brushlands (2.30 Mha) store 81 M tons of carbon and sequester 6.4 M tons C/yr. Grassland areas (1.18 Mha) contain 54 M tons C, but accumulate no C because of burning. Tree plantations cover 0.6 Mha. They contain 25 million tons of C and isolate about 2.6 million tons C annually. Agroforestry farms (5.7 Mha) contain 13 million tons of C and sequester 13.7 million tons C per year.

The total carbon budget of mountain areas in the Philippines reveals that they are able to absorb about 78% of total CO_2 emissions of the country per annum.

Future land-use changes in the mountain areas are based on the Philippine Forestry Master Plan for 1990–2015 (Asian Development Bank, 1990). By the year 2015, it is estimated that forest lands would store 775 M tons of carbon and could isolate 31.8 M tons C/yr. There will be 12% decline in total C storage relative to 1995 figures, but there will be a 17% increase in annual carbon sequestration as a result of an increase in total plantations.

The last part of the paper presents policy recommendations and possible courses of action by government and private entities to maximize the role of mountain lands in the mitigation of climate change. It is recommended that a carbon-offset programme be considered for the Philippines forest lands. Private corporations could be required to absorb fully or partially, through forest biomass, the amount of C they emit. The following issues and concerns must be addressed to make this operational:

- the carbon-sequestration rate of forest cover in the Philippines;
- a scheme of allocating carbon credits;
- value-added by the project; and
- monitoring.

REFERENCES

Asian Development Bank (1990) *Master Plan for Forestry Development.* Asian Development Bank, Manila, Philippines.

Forest Management Bureau (1996) *Philippines: Philippine Forestry Statistics.* Forest Management Bureau, DENR. Quezon City.

Lasco, R.D. (1997) Management of Philippine tropical forests: implications to global warming. Paper presented at the 8th Global Warming Conference. Columbia University, New York, USA. May 1997.

P.S. Ramakrishnan
School of Environmental Sciences
Jawaharlal Nehru University, New Delhi 110067
India

Ecological and human dimensions of 'global change' research

Traditional upland societies in the Asian tropics are closely linked with a variety of natural forested ecosystem types, ranging from dry deciduous to humid rainforest systems. They live in harmony with their natural environment, obtaining a variety of resources from the forest, such as food, fodder, fuelwood, medicinal plants, timber and other resources that may provide cash income. They are also involved with a wide range of land-use activities, chiefly for food production, from shifting agriculture on the one extreme to a variety of sedentary agro-ecosystem types on the other. All these agro-ecosystem types have close interconnections both with natural forest ecosystems and with complex village ecosystems. The various land-use types arising out of the livelihood activities of these traditional societies are integral components of a landscape mosaic. A characteristic feature of the landscape units lies in the fact that, traditionally, humans are well integrated into the ecosystem function, rather than the conventional way in which ecologists perceive humans in the more modern world as exploitative and sitting outside the ecosystem boundary. Furthermore, the biophysical complexity in the mountain landscape is compounded by extreme cultural diversity.

As agriculture and forestry-related activities are interlinked for many traditional societies in the region, natural ecosystems are subject to a variety of perturbation regimes. Traditionally, these perturbations are of small magnitude and infrequent. However, in more recent times, partly because of increased population pressure and partly due to the impact of industrial societies from outside the region where the traditional societies live, both the frequency and the intensity of perturbations have assumed alarming dimensions. They have led to large-scale deforestation, weed invasion by native and alien species, arrested succession of bamboo forests, site desertification and social disruptions. Arising out of these large-scale land transformations and/or linked to other factors that may be ecological, social, economic or cultural, a variety of complex agro-ecosystem types have evolved and are still evolving, in different parts of the Asian tropics in general and the Himalayan mountain region in particular, where traditional societies live.

Complex agro-ecosystems

Classifying complex agricultural systems and relating them to a gradient in management intensification is a difficult task. However, a grouping based loosely on the intensity of land-use and management, starting with shifting agriculture on the one extreme to high-energy-input modern agriculture on the other, provides a useful framework for discussing the relationship between agro-ecosystem complexity and function.

It is generally acknowledged that ecosystem complexity decreases as habitats change from forest to traditional agriculture and further to modern agriculture. If we plot total biodiversity (not just that obviously associated with the production system itself) at each point along this gradient, it is highly probable that the resulting relationship will be monotonic and decreasing, the exact form of the curve being somewhat uncertain. Available evidence on traditional agricultural systems, however, tends to suggest that the biodiversity dips sharply somewhere in the middle intensities of management. This is significant from the point of view of management of agricultural biodiversity with concerns for built-in agro-ecosystem stability, resilience and overall sustainability with optimal economic returns.

In a conceptual sense, for the traditional societies, there could be three different pathways for sustainable agriculture: (a) evolution by incremental change, (b) restoration through the contour pathway, and (c) development through the auto-route (i.e., 'modern' agriculture, based on heavy external energy subsidies). Realising that biodiversity does contribute in a variety of ways to ecosystem functions and that agro-ecosystems do harbour a great deal of biodiversity valuable for human welfare, it is reasonable that we aim for a mosaic of natural ecosystems coexisting with a wide variety of agro-ecosystem models derived through all the three pathways. Such a highly diversified landscape unit is likely to have a wide range of ecological niches conducive to enhancing biodiversity and, at the same time, to ensure sustainability of the managed landscape.

Managing the forested landscape

Attempts to manage tropical forests and harvest valuable species have not always been successful. Ecological inputs are important for determining management decisions. Knowledge from areas such as tree biology and architecture, patch dynamics, eco-physiology of developing forest communities, reproductive biology and nutrient cycling processes could all be integrated into the current management process and future management options. In such an integrated approach to management, the socio-economic and socio-cultural issues and traditional knowledge coming from the local communities need to be reconciled. Such an integration of eco-technologies is seen from our on-going studies in the Himalayan region, where the sustainability criterion has been the touchstone for designing management strategies.

Mobilising the local community in forest restoration and catchment protection, rainwater harvesting and its distribution, and in a variety of related eco-developmental works arising out of watershed management (e.g., agriculture, agroforestry, horticulture, animal husbandry, bamboo plantation and bamboo-based artisanal activities) has been carried out in the Himalayan and sub- Himalayan tracts of India by the author and his colleagues. Local involvement was made possible through direct interaction with villagers, through NGOs or through organised village-level societies.

Conclusions

Sustainable development of local communities, effective management of natural resources with concerns for conserving biodiversity, and indeed rehabilitation of degraded/altered ecosystems in the context of global change phenomena are all closely interlinked. Ecological issues are tied up with social, economic, anthropological and cultural dimensions, since the guiding principles of sustainable development cut across these disciplinary realms, with obvious trade-offs. This implies that we have to make a series of compromises to achieve sustainable development in such a way that we do not lose track of the ultimate objective: rehabilitation and management of natural resources in a manner that satisfies current needs, at the same time allowing for a variety of options for the future.

Though an ecosystem type (man-made ecosystems such as agriculture, a fish pond in a village, or a village itself visualised as an ecosystem; or natural ecosystems such as grazing land, forest or river) may be the appropriate unit for convenient handling of the issues involved in sustainable management of natural resources, a cluster of interacting ecosystem types (a 'landscape') may be the most effective for a holistic treatment. A watershed is one such landscape unit.

Further, from a sustainable developmental point of view, while one may bear in mind a long-term ideal objective to be achieved, ecological, social economic or cultural constraints may necessitate designing short-term strategies, for enabling peoples' participation in the developmental process. To quote one example, while forest-based economic activities and cash-crop plantation programme may be the most appropriate as a long-term alternative to shifting agriculture in north-east India, there is no option except to have a redeveloped agro-ecosystem package for the region, using traditional knowledge and technology as the starting point for a short-term strategy. The long-term strategy has to be reconciled with short-term requirements. Thus, sustainable development has spatial and temporal dimensions that need to be reconciled.

Indicators of sustainable development are varied. Therefore, here again, compromises are called for. Monitoring and evaluation has to be done using a number of diverse indicators. These may be:

- ecological (land-use changes, biomass quality and quantity, water quality and quantity, soil fertility, and energy efficiency);
- economic (monetary output/input analysis, capital savings or asset accumulation, and dependency ratio);
- social (quality of life with more easily measurable indicators such as health and hygiene, nutrition, food security, morbidity symptoms; the difficult to

quantify measures such as societal empowerment, and the less tangible ones in the area of social and cultural values).

We have considered all these diverse indicators to arrive at a meaningful sustainable developmental strategy for north-eastern hill areas of India.

Rehabilitation/sustainable development with peoples' participation demand closer interaction between ecologists and social scientists who have traditionally worked in isolation, using different paradigms for development. It also calls for interaction between developmental planners and the local communities that could trigger peoples' participation. In order to achieve this, developmental strategies have to be based on a value system that people can understand and appreciate, and therefore participate in the developmental process itself.

Claude Saint-Pierre
CADE (Consultants in Agriculture, Development and Environment)
20 bis, rue Fontaine St. Berthomieu, Agropolis, 34070 Montpellier
France

Building a shared view of future land-use in a project area: terraced agroecosystems in China and the Philippines

Farmers in terraced agroecosystems carry out a strategy of diversification on terraces, with the development of tree or cash crops, and a concentration of labour on terraces. Stakeholder analysis of land-use strategies shows consensus on terrace expansion in China and on terrace rehabilitation in the Philippines, while swidden cultivation on slopes, and crop diversification on terraces are areas of conflict between stakeholders. Development projects in these areas can focus their actions related to land-use on areas of consensus, and play a mediation role in areas of conflict.

Framework of analysis

The Daba mountains form the limit between Shaanxi and Sichuan provinces in China (32°N), while the Ifugao mountains are located in the south of the Central Cordillera, Luzon, in the Philippines (17°N). In these terrace-centred agro-ecosystems, the dichotomy between terraces and slopes is used as an entry point to analyse stakeholders' land-use strategies.

Land-use change and stakeholder analysis: Daba mountains

In the Daba mountains, the Han population is expanding terraces. Terraces dramatically increase labour productivity, particularly where women are performing farm operations while men leave for off-farm employment. They allow women to devote more time to animal husbandry and tree crops. One widely grown medicinal cash tree is *Eucommia*. Smallholders only plant *Eucommia* on the terraces, not on slopes. A few larger-scale farmers plant *Eucommia* on slopes; these are farmer-entrepreneurs able to secure access to land and credit.

A national scheme helps villages to purchase explosives, cement, and skilled masonry work to build terraces. *Eucommia* is an important species promoted by the Ministry of Forestry in South China. The Ministry, which is responsible for the afforestation of land with slopes above 25°, carries out detailed plantation design and provides seedlings. This planning system requires a minimum plantation size of several hectares.

Land-use change and stakeholder analysis: Ifugao mountains

In the Ifugao mountains, around small towns, due to emigration, off-farm work linked to tourism, and cultural change in the community, the Ifugao people are spending less time on terrace maintenance. Furthermore, an earthquake in 1990 caused the collapse of some terraces. There is a trend to terrace degradation, counterbalanced by the preservation of the traditional *muyong* woodlots at the head of terrace clusters, to protect water sources. Village communities are, however, keen to seize funding opportunities for concrete lining to irrigation canals.

The Ifugao mountains are adjacent to Benguet province which has developed into a major upland vegetable-producing area, and many young Ifugao have found jobs there. Migrants returning to the Ifugao mountains rent the degrading paddy terraces to grow vegetables.

Agencies, which include the Department of Environment and Natural Resources, NGOs and the Department of Agriculture, concentrate their activities in Benguet province, and have little stakeholder involvement in land-use change in the Ifugao region. The Ifugao Terraces Commission aims to preserve what is today the main asset of the Ifugao mountains for the tourism industry. The Commission recommends a zoning approach to distinguish the terraces to be maintained as rice terraces from those which can sustain multiple uses, including vegetable farming.

Table 1: Areas of consensus and conflict between stakeholders

	Consensus	Conflict
Daba Mountains	• Terrace expansion	• Agroforestry on terraces
		• Scale of tree plantations
		• Swiddening
Ifugao Mountains	• Control of terrace degradation	• Paddy field owners versus vegetable producers
	• Woodlot preservation	• Swiddening

Project land-use change strategies

In the absence of a specific land-use change strategy, the projects would have devoted part of their resources to activities directly aimed at reducing slope degradation. They might also have addressed as a priority the needs of the more active categories of project clients, who are the forestry farmer-entrepreneurs in the Daba mountains and the specialised vegetable producers in the Ifugao mountains. Stakeholder analysis carried out early during project preparation makes it possible to direct project resources to large numbers of smallholders, while supporting land-use systems with enhanced sustainability.

Table 2: Proposed project land-use change strategies

Focus on terraces, not on slope degradation		Mediation on diversification on terraces	
Daba	*Ifugao*	*Daba*	*Ifugao*
• Terracing	• Infrastructure rehabilitation	• Agroforestry • Smallholder forestry	• Terrace zoning • Technology for vegetables and paddy e.g., irrigation
Increased incomes, thus reduced slope degradation		*More forestry beneficiaries*	

The information originates from consultancies carried out for two projects: (1) Qinba Poverty Reduction Project of China's Leading Group for Poverty Reduction, funded by the World Bank, and (2) Central Cordillera Agricultural Project (Phase II) of the Department of Agriculture of the Philippines, funded by the European Union. The opinions in this paper are the author's alone.

Klaus Seeland
Chair of Forest Policy and Forest Economics, Department of Forest & Wood Sciences
Swiss Federal Institute of Technology, ETH-Zentrum, 8092 Zurich
Switzerland

The National Park management regime in Bhutan: historical background and problems

Context and historical background

The Kingdom of Bhutan is situated on the southern slopes of the Eastern Himalayas, covering an area of 40,076 km² and spanning various climatic zones, from sub-tropical to arctic-alpine. Southern Bhutan lies at an altitude of about 200 m, and the country rises to an altitude of 7,500 m in the north, where it constitutes a part of the Great Himalayan Range. The country is renowned as one of the world's hotspots of biodiversity, with more than 60% of the species endemic to the Eastern Himalayas. The vast majority of the population of 640,000 are Buddhists practising Lamaism. Bhutan has a Five Year Plan economy and a per capita income of $US478 in 1996, which places it among the least developed countries in the world (UNDP, 1996).

Bhutan has 72.5% forest cover; only 7.8% is arable land under cultivation. By 1997, approximately 26.5% of Bhutanese territory had been declared protected areas (RGOB, 1996). The establishment of protected areas as wildlife sanctuaries dates back to 1966, when the royal family's former hunting reserves in the south were to become the Phipsoo and Manas Wildlife Sanctuaries, by Royal Government order. The Bhutan Forest Act of 1969 (RGOB, 1969) adopted the concept of Reserved Forests from India, and all forest became state forest in which many restrictions limited the formerly free access of the users to timber and non-timber forest products. After the introduction of forest legislation, the World Wildlife Fund (USA) sought the approval of some of Bhutan's top officials for a national scheme to establish wildlife sanctuaries, parks and forest reserves according to the US model. Settlements inside the park area became one of the major issues for conflict between the state authorities and the local population, whose customary rights to use the natural resources of their environment, including game, were drastically reduced without any compensation by the state.

National parks, wildlife sanctuaries and forest reserves could only be established and maintained with foreign aid. Financial gains from the revenue drawn from state forests may have begun to play a role after the first formulation of a forest policy in 1974, when the country's socio-economic development put increasing strain on the Government's financial resources. In 1974, the Government of Bhutan issued a notification on the Creation of

Wildlife-Sanctuaries/Parks/Forest-Reserves, geographically demarcating eight protected areas, chiefly on the basis of a random assessment. The ranking of National Park, Wildlife Sanctuary, Strict Nature Reserve and Reserved Forests reflects the order of importance given to the respective site.

Achievements over the last decade

A process to revise the Government's Nature Conservation Policy was undertaken from 1989 to 1991. Until 1993, there was no National Conservation Strategy. Together with suggested amendments, a National Forest Policy was drafted in 1985. In 1993, a further notification was issued by the RGOB, whereby the protected areas were revised in order to rationalise the biodiversity representation and protection needs. Through this notification, the new concept of buffer zones was introduced to some protected areas, so that the effective area under protection was increased to 10,513 km²: about 1,000 km² more than the area demarcated in the 1984 notification. Compared to the first demarcation in 1974, there was an increment of 13.1%.

Problems and Prospects

Activities threatening Bhutan's natural wealth are perpetrated by different sections of Bhutanese society who engage in illegal timber-felling and the poaching of endangered species. Intruders, including Tibetan gangs who cross the border to collect highly valuable mushrooms and medicinal herbs, have become a nuisance to the Northern Wildlife Circle. Similarly, political activists and militant guerrilla fighters are intruding into the southern national parks in order to escape capture by the Indian Army.

The major issues to be tackled by the Nature Conservation Section include crop damage by the increasing number of wildlife inside and outside protected areas, and wildlife attacking domestic animals and villagers. Another important task is to harmonise conservation objectives and local peoples' daily requirements of timber and non-timber forest products (RGOB, 1995). The conflicts that resulted in the years after certain areas were declared "protected", led to a management policy based on close cooperation with the people inhabiting the protected areas, but not to any formal institutional participation. In this respect, a clear policy decision on the national park management regime must be taken.

REFERENCES

Royal Government of Bhutan (RGOB) (1969) The Bhutan Forest Act, Ministry of Trade, Industries & Forests, Thimphu.

Royal Government of Bhutan (RGOB) (1995) Forest and Nature Conservation Act 1995, Thimphu.

Royal Government of Bhutan (RGOB) (1996) Eighth Five Year Plan (1997–2002), vol. I, Main Document, Ministry of Planning, Thimphu.

United Nations Development Programme (UNDP) (1996) Development Cooperation Report, 1994, UNDP Bhutan, Thimphu.

Joëlle Smadja
CNRS, UPR 299, "Milieux, Sociétés & Cultures en Himalaya"
1 place Aristide Briand, 92195 Meudon Bellevue
France.

Natural resources and land-use changes in Nepal: villagers' initiatives to replace forest resources

Trees are one of the primary resources used by the peasants of Nepal – they provide fodder, firewood and timber – and are also linked to slope stability, water availability, and other issues. The regression of the forest over the last decades has led to alarming hypotheses concerning the perpetuity of Himalayan agrarian systems (Eckholm, 1975; Caine and Mool, 1981). However, recent studies emphasize that we should be very prudent in how we interpret what we observe (cf. Ives and Messerli, 1989). These observations make no sense unless they are situated within a historical and cultural context. This is the main thrust of the interdisciplinary programme, "Explanations of the diversity and the evolution of some landscapes in Himalaya: Examples in Nepal and Ladakh", which has been undertaken by the CNRS UPR 299 laboratory. Research focuses on the physical features which can explain diverse landscapes; archival documents; and qualitative and quantitative field enquiries related to the perception of space and resource management, land-use, tenure, taxation, and other topics.

From our results it is clear that: 1) deforestation in the Himalaya is not a new environmental problem, and 2) in places where the forest is no longer available, strategies have been developed to provide fodder and wood.

Deforestation is not a new environmental problem

Two texts illustrate that deforestation is not a new environmental problem. The first text concerning forest or tree protection in Nepal is not at all recent. It is the fourteenth edict from Ram Shah, king of Gorkha between 1606 and 1636.

> "*Forests are to be preserved near watering places. If there are no trees, there will be no water whenever one looks for it. The watering places will become dry. If forests are cut down, there will be landslides. If there are many landslides, there will be great destruction. Accidents also destroy the fields. Without forests, the householders' work cannot be accomplished. Therefore, he who cuts down the forest near a watering place will be fined five rupees.*" (after Riccardi, 1977).

Other texts confirming the above have been recorded by Ramirez (1996) within the framework of the programme.

A second illustration is a text by Dr W. Hoffmeister (1848), a travelling physician who crossed Nepal in 1845. He wrote:

> "*We followed, for some time, the ridge of this pass, proceeding in a north-westerly direction, and thus enjoyed an opportunity of observing the marked difference between the north and the south and east sides. The two last are bare and treeless, while the former is clothed with noble forests.*"

Other botanists during the same period confirmed this contrast. The usual interpretation nowadays is that these southern slopes have been deforested because of human pressure on the land.

Thus, these two texts point to the facts that :

- in these fragile mountains, cutting down trees is not a new environmental problem: people have been concerned with tree protection for a long time;
- if deforestation is a real problem in some areas of Nepal, the observations made today are not always related to demographic growth since the 1950s, nor to recent human pressure on the land.

Strategies related to fodder and wood shortage

Three examples collected during our fieldwork from 1979 to 1996 illustrate strategies related to fodder and wood shortage.

In the middle mountains of Central Nepal (in Nuwakot District) the upper part of the slope (above 2,000 m), where only wheat was cultivated until 1988, had been converted into irrigated rice fields. A new variety of red rice has been introduced. With its long straw it provides, besides other benefits, more fodder for livestock and thus in part compensates for the degradation of shrub and forests, which are now protected. This change was hardly predictable when researchers worked in this area between 1979 and 1985.

In many places in Nepal, rotten shingles or thatched roofs have been replaced with recycled tins, corrugated iron roofing or wherever possible, by slate roofs. These changes in roofing materials alleviate the strain upon forest resources and in these regions, thatching material is now used mainly for cattle fodder.

A final example is that, with the decrease of forest resources, trees have become increasingly numerous in the fields. In many areas, the forest is no longer used as a resource, as trees from the cultivated areas provide for all wood and fodder needs (Carter, 1992; Gilmour, 1988; Smadja, 1995).

From the beginning of our programme, we have emphasised that landscapes in Nepal are characterised by continuous adaptations and innovations. But all changes do not ineluctably lead to degradation (cf. Fox, 1993). And when we talk about deforestation in Nepal, we must also consider the numerous trees in the fields and the different strategies people adopt to meet their needs.

REFERENCES

Caine, N., Mool, P.K. (1981) Channel geometry and flow estimates for two small mountain streams in the Middle Hills, Nepal. *Mountain Research and Development*, **1** (3–4): 231–243

Carter, E.J. (1992) Tree cultivation on private land in the Middle Hills of Nepal: lessons from some villagers of Dolakha District. *Mountain Research and Development*, **12** (3): 241–255

Eckholm, E. P. (1975) The deterioration of mountain environments. *Science*, **189**: 764–770

Fox, J. (1993) Forest resources in a Nepali Village in 1980 and 1990: the positive influence of population growth. *Mountain Research and Development*, **13** (1): 89–98

Gilmour, D.A. (1988) Not seeing the trees for the forest: a reappraisal of the deforestation crisis in two hill districts of Nepal. *Mountain Research and Development*, **8** (4): 343–350

Hoffmeister, W. (1848) Travels in Ceylon and continental India. Including Nepal and other parts of the Himalayas, to the borders of Tibet. Appendices addressed to baron Von Humboldt on the vegetation of Himalaya. William P. Kennedy, Edinburgh.

Ives, J.D. and Messerli, B. (1989) *The Himalayan Dilemma. Reconciling Development and Conservation.* London and New York, The United Nation University, Routledge.

Ramirez, P. (1996) Le rôle de l'Etat dans la transformation des paysages. Internal document. CNRS Meudon France.

Riccardi, T. (1977) The Royal Edicts of King Rama Shah of Gorkha. *Kailash*, **5** (1)

Smadja, J. (1995) Sur une dégradation annoncée des milieux népalais: initiatives villageoises pour remplacer les ressources forestières. *Natures Sciences Sociétés*, **3** (3): 190–204.

Luis Daniel Llambí
Universidad de los Andes, Centro de Investigaciones Ecologicas de los Andes Tropicales
Facultad de Cienias, La Hechicera, Mérida
Venezuela

Introduction: Latin American mountains

Latin America holds a very particular position in the discussion over the world's mountains, with the Andes being the backbone for the entire South American continent. As the longest cordillera on Earth, stretching from the Equator to the Antarctic, the Andes comprise an immense diversity of ecosystems, cultural traditions and livelihood strategies. This is why we consider the first key issue for the region the analysis of the opportunities offered by this cultural and ecological megadiversity. One of its most important aspects in terms of its value for the food security of its people is the vast diversity of traditional agricultural ecotechnologies. An evaluation of their sustainability, productivity and environmental impact is a clear research priority for the region.

The papers by Preston and Llambí and Sarmiento analyse from this perspective two case studies from the Southern Bolivian and Venezuelan Andes. Preston focuses on the assessment of the erosive impact of an important productive strategy in the Andes: agropastoralism. Although there are some signs of erosion taking place in the area, Preston indicates that the present land-use strategy, which involves complex patterns of cattle movement, allows the utilisation of different environments and the maintenance of a sustainable system with low-erosive impact. In contrast, Llambí and Sarmiento, analysing the long fallows practised in a traditional agroecosystem of the Venezuelan High Andes, question the sustainability of this system in terms of its successional capacity to restore the original vegetation structure and appropriate levels for a series of key soil parameters, such as microbial biomass and soil organic matter. Paradoxically, while the Bolivian system had long been assumed to be degradative because of "foolish" farm practices, the Venezuelan fallow agriculture had been assumed to be highly sustainable. These case studies clearly show the importance of detailed agroecological studies before reaching any conclusion about the sustainability of a particular agroecosystem. An important lacuna of past and present research identified in the workshop concerned the integration in these studies of the socio-economic dimensions of the Latin American rural reality. Complex issues such as the participation of local communities in research, the social impact of drugs production in the region and the integration of Latin American rural economies in the emerging globalised markets are just some examples of the problems that need to be tackled in the near future.

Another key role played by mountains is the maintenance of water resources. This is particularly important in the case of the Andes both because of their connection with the Amazon Basin and since a big proportion of the Latin American metropolises such as Santiago, Bogota, and La Paz directly depend on the water from Andean catchments. Schrott´s paper demonstrates the hydrological significance of high mountain permafrost, a widespread cryogenic phenomenon covering larger areas than the more visible and studied glaciers. This study stresses the possible hazardous impacts of climatic change through permafrost degradation. This could be one of the most important consequences of global change in a region which, in sharp contrast with the situation in the First World, does not conceives global change as a fundamental priority for research.

The environmental impact of tourism is another issue which has received far more attention in other mountain areas of the world like the Himalayas, and is starting to be considered in the Andean context, as shown by Shackley´s paper. Her analysis of the adverse impact of trekking tourism on the Inca Trail to Machu Picchu pinpoints the need for appropriate strategies for tourist management in the face of rapidly increasing number of visitors. This is a situation that is likely to be repeated in many other mountain areas of Latin America.

To sum up, the big challenge ahead of us is to help decision makers understand the immense opportunities that the Andes offer for fostering Latin America's development, and to devise sustainable and socially just strategies to do so.

Luis Daniel Llambí and Lina Sarmiento
Centro de Investigaciones Ecológicas de los Andes Tropicales
Universidad de los Andes, Facultad de Ciencias, La Hechicera, Mérida
Venezuela

Ecosystem restoration during the long fallow periods in the traditional potato agriculture of the Venezuelan High Andes

In the Venezuelan High Andes (3,300 to 3,800 m), the traditional agricultural system of potato cultivation alternates short periods of production with long fallows (five or more years). During the fallow, a secondary succession occurs, allowing restoration of soil fertility and vegetation. Even though the process has been the subject of a number of studies in the Bolivian "punas" and the Colombian and Venezuelan "paramos", the ecological mechanisms of fertility recovery and the ecosystem impacts of this agriculture are not well understood (Hervé, 1994). The analysis of long-fallow agriculture with an ecosystem perspective has particular importance, since a substantial number of the Andean traditional systems are inside national parks and are suffering a recent process of transformation to more intensive and harmful land-use strategies.

In this study, our initial hypotheses were: (1) cultivation of paramo areas produces a significant decrease of soil microbial biomass; (2) the fallow period allows for the recovery of the diversity and biovolume of the plant community and soil microbial biomass; (3) these changes can be related to the restoration of the effectiveness of N-cycling and the recovery of fertility.

To analyse the dynamics of soil and vegetation restoration, we compared 36 successional fields with different ages (one to nine years, four replicates per year) and four areas of the never-cultivated paramo ecosystem. The study was undertaken during the wet season of 1996 in Gavidia, a valley of glacial origin with about 500 inhabitants, in Venezuela's Sierra Nevada National Park. A composite sample was obtained for each plot, and the following soil parameters were determined: soil total C and N; microbial biomass N (fumigation-extraction method: Brookes et al., 1985); NO_3; NH_4; P; Ca; Mg; Na; K; pH; cation exchange capacity; and texture. The vegetation was characterized using the point quadrat method (100 points per plot), and its total biovolume, species richness and diversity (Shannon-Wiener index) were determined.

The results show a very high heterogeneity between plots for all soil parameters. Principal Component Analysis shows that this heterogeneity can be partially associated with the geomorphological position of the plots. A successional recovery of the soil organic matter or of any of the nutrients evaluated is not observed, a result which has also been obtained in previous studies on traditional fallow agriculture in the tropical Andes (Sarmiento and

Monasterio, 1993; Ferwerda, 1987; Hervé, 1994). We did not observe an increase in microbial biomass, a result which contradicts our original hypothesis and the observations of other studies in temperate agro-ecosystems (Insam and Hasselwandder, 1989; Insam and Domsch, 1988; Sanstruckova, 1992). So, it is possible that processes not considered here, such as N-mineralisation dynamics or nitrification potentials, could explain the unresolved problem of fertility restoration during the fallow. However, our results have to be interpreted in the light of the high spatial heterogeneity, which could be partially obscuring real successional tendencies. We believe that this study shows the paramount importance of considering the high geomorphologic and edaphic heterogeneity which characterises these environments in the study and management of fertility in high mountain agro-ecosystems.

Analyzing vegetation dynamics, we observe that there is no successional increase in plant biovolume. in the case of species richness, there is an initial increase from the first to the second year, and then no apparent change up to the ninth year. Only species diversity increases steadily throughout succession.

Comparing the natural ecosystem with the successional plots, we detect that agricultural use causes a statistically significant decrease in microbial biomass N, which is more than twice as high in the never-cultivated plots, suggesting that this parameter is an excellent indicator of agricultural disturbance on paramo ecosystems. The use of microbial biomass as a disturbance indicator is supported by several authors, such as Gregorich *et al.* (1995); similar results have been obtained by the Tropical Soil Biology and Fertility Program in a number of contrasting savanna and forest sites of the lowland tropics (Woomer *et al.*, 1994). We also detect a statistically significant decrease in plant biovolume, species richness and diversity (Kruskall-Wallis, $\alpha=0.05$) and decreases in total soil C and N, pH, Ca and Mg which, however, are not statistically significant.

So our results show that agricultural disturbance on the paramo ecosystem causes a series of negative impacts on key soil properties, which are not restored in the fallow period currently used by the farmers. This questions the sustainability and the assumed high conservationist value of these traditional agro-ecosystems. However, the observed successional increase in diversity, and the fact that a good number of the successional species are not present in the original ecosystem, indicate that the fallow allows the maintenance of a higher landscape diversity than an intensive monoculture or the undisturbed ecosystem alone would allow.

REFERENCES

Brookes, P.C., Landman, A., Pruden, G. and Jenkinson. D.S. (1985) Chloroform fumigation and the release of soil nitrogen: a rapid direct extraction method to measure microbial biomass nitrogen from soil. *Soil Biology and Biochemistry*, **17** (6): 837–842.

Ferwerda, W. (1987) The influence of potato cultivation on the natural bunchgrass paramo in the Colombian Cordillera Oriental, Internal Report. No. 220. Hugo de Vries Laboratory, University of Amsterdam.

Gregorich, E.G., Carter, M.R., Angers, D.A., Monreal, C.M. and Ellert., B.H. (1995) Towards a minimum data set to assess soil organic matter quality in agricultural soils. *Canadian Journal of Soil Science*, **74**: 367–385.

Hervé, D. (1994) Respuesta de los componentes de la fertilidad del suelo a la duración del descanso. In Hervé, D., Genin, D. and Riviere G. (eds), *Dinámicas del Descanso de la Tierra en los Andes.* IBTA-ORSTOM, La Paz. pp. 15–36.

Insam, H. and Domsch, K.H. (1988) Relationship between soil organic carbon and microbial biomass on a chronosequence of reclamation sites. *Microbial Ecology*, **15**: 177–188.

Insam, H. and Hasselwandter, K. (1989) Metabolic quotient of soil microflora in relation to plant succession. *Oecologia* (Berlin), **79**: 174–178.

Santruckova, H. (1992) Microbial biomass activity and soil respiration in relation to secondary succession. *Pedobiologia*, **36**: 341–350.

Sarmiento, L. and Monasterio, M. (1993) Elementos para la interpretación ecológica de un sistema agrícola campesino de los páramos Venezolanos. In Rabey, M. (ed.), *El Uso Tradicional de los Recursos Naturales en Montañas: Tradición y Transformación.* UNESCO-ORCYT, Montevideo-Uruguay.

Woomer, P.L., Martin, A., Albrecht, R., Resck, D.V. and Scharpenseel. H.W. (1994) The importance and management of soil organic matter in the tropics. In Woomer, P.L. and Swift. M.J. (eds.), *The Biological Management of Tropical Soil Fertility.* John Wiley and Sons, Chichester.

M. A. McDonald[1], P.A. Stevens[2], J.R. Healey[1] and P.V. Devi Prasad[3]
[1]School of Agricultural and Forest Sciences, University of Wales
Bangor, United Kingdom
[2]Institute of Terrestrial Ecology, Bangor, United Kingdom
[3]Department of Life Sciences, University of the West Indies, Mona, Kingston, Jamaica

Maintenance of soil fertility on steeplands in the Blue Mountains of Jamaica: the role of contour hedgerows

Introduction

The context for this study is the urgent need to find a stable and sustainable alternative to shifting cultivation on steeplands in the tropics. In Jamaica, as in other mountainous regions, it has been accepted that cultivation will continue on many areas of sloping land, and ways must be found to make this environmentally acceptable. Suitable measures must provide smallholder farmers with the means to sustain crop yields and reduce labour inputs. The principal objective of the work reported here was to investigate the consequences of forest clearance by shifting cultivators on soil fertility. The use of a potential agroforestry system – contour hedgerows – for soil conservation was also investigated.

Methods

An experiment was undertaken with the participation of local farmers in the watershed of the Green River, a head-water tributary of the Yallahs River which is the main supplier of domestic water to the municipality of Kingston, Jamaica. The elevation is around 1500 m; annual rainfall averages 2500 mm.
Experimental plots were established in:

1. Secondary forest;
2. Forest cleared, burned and subsequently maintained weed-free;
3. Forest cleared, burned and planted with agricultural crops;
4. Forest cleared, burned and planted with agricultural crops intercropped with hedgerows of *Calliandra calothyrsus* – a ubiquitous, locally popular species.

Four blocks each containing one plot of each treatment were established in areas of secondary forest, between 20–30° slope; each block was managed by a participating local farmer. A mixture of the major crops cultivated in the area for subsistence and local markets were established in the agricultural plots. The hedgerow system (three per plot at 5 m spacing) was designed after Young (1989). The hedges were cut bi-annually since planting to a height of about 30cm, and never allowed to grow more than about 1 m tall, at which point they started to shade the crops.

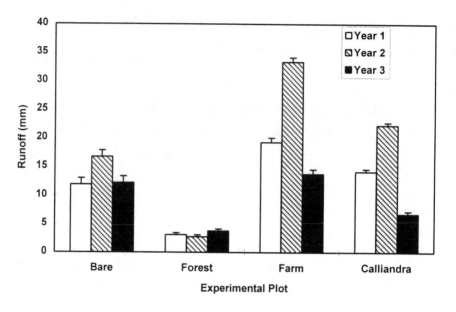

Figure 1. Runoff from the experimental plots, expressed as a proportion of annual rainfall

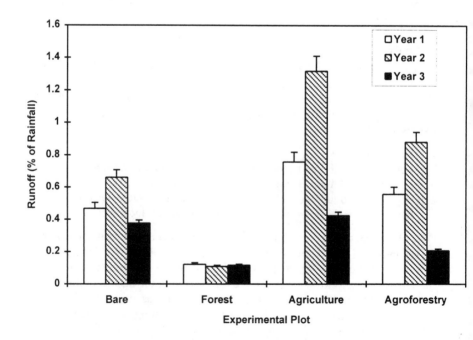

Figure 2. Sediments eroded from the experimental plots.

Samples of runoff and sediment were collected on a fortnightly basis from September 1992 to September 1995, and after every rain event >25 mm from March 1996 to present. *In-situ* incubations (30 day) were conducted in the plots in October from 1994–1997 to assess rates of total nitrogen mineralisation and nitrification.

Results

Levels of runoff over the course of each year were very low in all plots (Figure 1). Individual storms accounted for large proportions of total runoff. Consequently, measurements of associated erosion were much lower than previously published estimates (e.g. McGregor, 1988). However, forest clearance followed by cultivation resulted in a 31-fold increase in the quantity of eroded sediments (Figure 2). In the agroforestry plots, erosion was up to 30% less than for the conventionally farmed plots in the first year after clearance. Rates of erosion increased significantly over time in the bare treatment plots, whilst they dropped significantly in the agriculture and agroforestry plots, and showed little change in the forest plots.

This pattern was mirrored in the rates of loss of mineral nutrients. In the agroforestry plots, nutrient losses should have been more than offset by the productivity of the hedgerows. However, although there was an increase in the rate of nitrogen mineralisation rate below the hedgerows, there was net immobilisation in the farmed areas between the hedges. This was probably as a result of the prunings being scattered on the soil surface.

Discussion and conclusions

Considerable protection is offered to both the soil and water resource by the forest, even though it is of a secondary nature. In particular, the forest acts as a buffer against fluctuations in runoff associated with rainfall events, which may be of significance in large-scale events. Agricultural use of cleared land does result in increased erosion and a consequent reduction in soil fertility (leading to the low sustainability of the current cultivation practice). However, the erosion rates are not as high as anticipated from previous estimates, and the use of contour hedgerows reduces them to levels which may be considered acceptable (Morgan, 1986). Thus the barrier hedgerow system obviously has potential for sloping lands such as encountered in this study, but its adoption by farmers depends on the simplification of its establishment and its development into a flexible system (Njoroge and Rao, 1995).

Acknowledgements

This work has been funded by the Forestry Research Programme of the Department for International Development. However, DFID can accept no responsibility for any information given or views expressed.

REFERENCES

McGregor, D.F.M. (1988) An investigation of soil status and landuse on a steeply sloping hillside, Blue Mountains, Jamaica. *Singapore Journal of Tropical Geography,* **9**: 60–71.

Morgan, R.P.C. (1979) *Soil Erosion.* Longman, New York.

Njoroge, M. and Rao, M.R. (1995) *Barrier Hedgerow Intercropping for Soil and Water Conservation on Sloping Lands.* International Centre for Research in Agroforestry, Nairobi.

Young, A. (1989) *Agroforestry for Soil Conservation.* CAB International, UK.

David Preston[1], Mark Macklin[1] and Máximo Liberman[2]
[1]Geography Department, University of Leeds
Leeds, LS2 9JT, United Kingdom
[2]Instituto de Ecología, La Paz, Bolivia

Farmer strategies, production systems and environmental change: novel interpretations in the Southern Bolivian Andes

Research between 1992 and the present, funded by the STD programme of the Commission of the European Communities, in Tarija Department, southern Bolivia, has focused on the nature and processes of environmental change in relation to the survival strategies of rural households.

Research methods included monitoring a sample of 18 households in three localities over a three-year period, to study their livelihoods, responses to natural and socio-economic environmental stresses, and perceptions of past and present activities in relation to the natural environment. Natural scientists studied geomorphology and soils to attempt to determine the chronology of landscape change over the past millennium.

Farming includes crop growing on valley bottoms, often with irrigation (rainfall is 400–1000 mm), largely for subsistence with maize, potatoes, etc., and livestock rearing. Livestock use a wide range of micro-environments for grazing: locally, regionally on high mountain slopes, and five days walk distant, on wetter east-facing hillslopes during the dry season (Preston, 1998).

Migration is crucial to the well-being of most households and has been for at least 50 years. This involves travel to nearby Argentina to work in areas of intensive commercial farming, for the sugar cane harvest and in vegetable and grape growing areas over a large part of northern Argentina. People also travel to the nearby town of Tarija (women for domestic service); to Santa Cruz, which is a centre of lowland commercial farming; and to the Bermejo sugar plantations on the Argentine border.

Conclusions

Crop production is sustainable on near-level valley bottoms and dry-farmed lower hillslopes (Preston *et al.*, 1997).
Livestock grazing is locally sustainable by the use of a wide range of micro-environments. Sheep numbers have declined and informants report less land-used for crops. The colonisation by Churqui (*Acacia caven*) bushes of areas used less intensively has taken place, with positive effects on soil quality.
Erosion – widely reported as covering 75 per cent of the central Tarija valleys – is largely historic and on areas of steeply-sloping and easily-eroded lacustrine and aeolian deposits. We estimate that only 10 per cent of surface area is subject

to active erosion. Denuded hillsides seem stable. A single optically-stimulated luminescence date suggests that the period of most active deposition dates back 100,000 years; lichenometric studies indicate that flood frequency and magnitude, in a sample sub-catchment, have declined in the past 100 years in response to secular changes in climate (Warburton *et al.*, 1997).

REFERENCES

Preston, D., Macklin, M and Warburton, J. (1997) Fewer people, less erosion: the 20th century in Southern Bolivia. *Geographical Journal*, **163**(2): 198–205.
Preston, D. (1998) Post-peasant capitalist graziers – the 21st century in Southern Bolivia. *Mountain Research and Development*, **18** (2) (In press).
Warburton, J., Macklin, M. and Preston, D. (1997) Fluvial hazards in a steepland mountain environment, Southern Bolivia. *Geojournal*, (In press).

Lothar Schrott
Geographisches Institut, Universität Bonn
Meckenheimer Allee 166, 53115 Bonn
Germany

The hydrological significance of permafrost in the semi-arid Andes

Introduction

In the semi-arid region of the Andes in western Argentina, economic possibilities are closely linked to the annual amount of meltwater. Agriculture, for example, totally depends on irrigation, mainly drawing upon the meltwater of the mountain rivers. In this part of the Andes, detailed analyses of the glacier and permafrost hydrology are very important because of water scarcity and increasing desertification.

Until recently, very little was known about the hydrology of periglacial environments in high mountain areas (Blumstengel and Harris, 1988; Gardner and Bajewsky, 1987; Barsch et al., 1994). Especially in semi-arid and arid mountains, the melting of frozen ground could play an important role in the water balance. However, the extent and hydrological significance of high mountain permafrost has yet to be demonstrated (Corte, 1976; Schrott 1996).

Study site and methods

The selected study area in western Argentina is situated at 30°S in the Cordillera Principal (province of San Juan) near the Chilean border. The upper catchment of the Agua Negra has a surface of 57 km² and represents an area typical of the semi-arid subtropical region, with small glaciers and large rock glaciers. Different approaches were applied in order to quantitatively estimate the potential water store of permafrost bodies and glaciers and to show how much meltwater they constitute throughout the ablation period. Permafrost surveys were carried out using a combination of direct observation (e.g., distribution of rock glaciers and debris frozen by ice lenses) and indirect methods (e.g., borehole temperature and seismic refraction sounding). On the basis of an inventory, taking account of surface and thickness, the potential ice content of glaciers and rock glaciers was calculated (see Table 1) (Schrott, 1991; 1994).

Results and discussion

The following estimate gives an idea of the probable total water stored in permafrost: given an ice content of rock glaciers of approximately 60% and using an average thickness of 50 m and the measured surface areas, the water content of the active rock glaciers was calculated to be 62×10^6 m³. Even without considering the other permafrost areas, this corresponds to about 70% of the estimated water volume of the glacier. This seems reasonable since the area of active rock glaciers is larger than the area of glaciers in the basin.

Table 1: Inventory of the surface of glacier ice, rock glacier and permafrost zones. Discharge and potential volumes of water storages in the upper Agua Negra catchment (57 km²)

	Thickness on average [m]	Surface [km²] / [%]	Ice content [%]	Ice/water volume [10^6 m³]
Active rock glacier	50	2.07 / 3.1	60	62
Glacier ice	50	1.78 / 3.6	100	89
Continuous permafrost	20–30	9 / 15.8	20	36–54
Discontinuous permafrost	20–30	44 / 84.2	20	44–66[a]
Total ice/water volume				231–271
Melting of frozen ground and meltwater from areas underlain by permafrost				1.04[b]
Measured discharge during the ablation period 1990/91				5.06

(a) based on estimated permafrost distribution of 25% of the surface
(b) estimation during the ablation period 1990/91

Using the discharge rate (5–8 l s⁻¹) of the largest rock glacier in the basin as a minimum value, and taking into account the meltwater supplied by all the other existing active rock glaciers in the catchment, the estimated discharge would constitute as much as about 13% of the mean overall streamflow for the summer months.

Moreover, this does not include the other potential water sources in the permafrost areas. Changes in the streamflow between the glacier tongue – situated at the head of the catchment – and 9 km downstream confirm that the melting of frozen ground constitutes an important part of the basin discharge. That means that, after snowmelt (November/December), the melting permafrost produces a discharge which significantly increases from the glacier tongue to the gauge station and constitutes about 30% of the total discharge. For the whole ablation period 1990/91, the permafrost melt constitutes about 20% of the total discharge. This rather simple estimation shows that the water storage of permafrost and the water supply due to melting processes plays an important role in the water balance, especially in semiarid or arid mountain regions (Trombotto *et al.*, 1997; Schrott, 1996).

Due to the fact that permafrost areas are highly sensitive to climate change, there is a possibility of permafrost degradation in this part of the Andes and, as a consequence, a shift of the lower limit of mountain permafrost due to a rise in its temperature. The loss of water resources and the potential destabilisation of slopes which were previously frozen will create a variety of hazards (e.g. debris flows) (Haeberli *et al.*, 1993; Koster, 1994).

REFERENCES

Barsch, D., Happoldt, H., Mäusbacher, R., Schrott, L. and Schukraft, G. (1994) Discharge and fluvial sediment transport in a semiarid catchment, Agua Negra, San Juan, Argentina, In Ergenzinger, P. and Schmidt, K.-H. (eds.), Dynamics and Geomorphology of Mountain Rivers. *Lecture Notes in Earth Sciences*, 52. Springer, Berlin. pp. 213–224.

Blumstengel, W. and Harris, S.A. (1988) Observations on an active lobate rock glacier. Slims River Valley, St. Elias Range, Canada. Fifth International Conference on Permafrost, Trondheim, Proceedings vol.1, pp. 689–694.

Corte, A.E. (1976) The hydrological significance of rock glaciers. *Journal of Glaciology*, 17, 75: 157–158.

Gardner, J.S. and Bajewsky, I. (1987) Hilda rock glacier stream discharge and sediment load characteristics, Sunwapta pass area, Canadian Rocky Mountains. In Giardino, J.R., Shroder, J.F.Jr. and Vitek, J.D. (eds.), *Rock Glaciers.* Allen & Unwin, Boston. pp. 161–175.

Haeberli, W., Guodong, C., Gorbunov, A. P. and Harris, S. A. (1993) Mountain permafrost and climatic change. *Permafrost and Periglacial Processes*, 4: 165–174.

Koster, E. A. (1994) Global warming and periglacial landscapes. In Roberts, N. (ed.), *The Changing Global Environment.* Blackwell, Oxford. pp. 127–150.

Schrott, L. (1991) Global solar radiation, soil temperature and permafrost in the central Andes, Argentina: a progress report. *Permafrost and Periglacial Processes*, 2: 59–66.

Schrott, L. (1994) Die Solarstrahlung als steuernder Faktor im Geosystem der subtropischen semiariden Hochanden (Agua Negra, San Juan, Argentinien). *Heidelberger Geographische Arbeiten*, Heft 94, Heidelberg.

Schrott, L. (1996) Some geomorphological-hydrological aspects of rock glaciers in the Andes (San Juan, Argentina). *Zeitschrift für Geomorphologie* N.F., Suppl.-Bd. 104: 161–173.

Trombotto, D., Buk, E. and Hernandez, J. (1997) Monitoring of mountain permafrost in the Central Andes, Cordon del Plata, Mendoza, Argentina. *Permafrost and Periglacial Processes*, 8: 123–129.

Myra Shackley
Centre for Tourism and Visitor Management, Business School
Nottingham Trent University, Nottingham NG1 4BU
United Kingdom

Trekking tourism in Peru: the case of the Inca Trail

The Himalayas are not the only mountain region experiencing problems with the management of trekking tourism. Newly-popular trekking routes in Africa, particularly in the Rwenzori range (Mountains of the Moon) and Mount Kilimanjaro are already becoming over-used and eroded. The expansion of special-interest tourism in South America has also included the development and promotion of mountain walking and trekking, with the most popular destinations being found in Peru and Chile. Visitor numbers on the Inca Trail to Machu Picchu in Peru, now one of the most famous trekking routes in the world, have increased by 800% over the past decade, causing serious problems with trail deterioration. These are compounded by poor and ineffective management, combined with commercial pressures from trekking agencies and local communities. Despite attempts to control visitor impact, both the fabric and atmosphere of the trail are disintegrating under sheer weight of numbers.

The 43 km Inca Trail to Machu Picchu in Peru follows part of one of the major Inca highways and has become the most heavily utilised trekking route in South America. Most hikers start from Km88 on the Cusco-Quillebamba railway and take three to four days to reach Machu Picchu, passing through a stunning combination of Inca ruins, magnificent views and ecological variety within the 325 km^2 of the Machu Picchu Historical Sanctuary. The Trail can only be walked in one direction, with the exception of a short stretch approaching Macchu Picchu. Although there are some tough climbs and three passes above 3,660 m, the trail attracts a number of inexperienced trekkers. The route was originally part of an Inca sacred way, and has become something of a contemporary pilgrimage as well. Visitors can walk the trail individually (permit cost $17), or in organised groups ($70–150) accompanied by guides and porters. A shorter additional Trail from Km104 has recently been opened (permit cost $12) starting at the ruins of Chachabamba and connecting with the main trail at Winay-Wayna.

Around 40,000 visitors per year now walk the Trail, together with large numbers of porters and guides. The resulting impacts include:

- trail damage from excess visitors and heavily-laden porters wearing out the steep masonry stairs and pathways;
- infamous litter and pollution problems, with no facilities for waste disposal;
- the lighting of campfires using forest resources, probably responsible for the serious fires of 1988, 1994 and 1997;

- damage to ruins by climbing and lighting fires;
- dirty and overcrowded campsites with few sanitary facilities;
- increasing problems with theft and mugging.

Problems are made worse by the lack of a coherent administration policy. Ten government agencies have some form of jurisdiction in the area, with none achieving overall authority. Rapidly-increasing numbers of visitors have destroyed the original spirit of place, and the nature of the existing visitor experience is often recorded as unsatisfactory. Pressure to increase visitor numbers still further comes from commercial trekking agencies in Cusco and local communities employed as porters.

Previous ineffective attempts to control tourism impacts have included:

- enforcement of a one-way system;
- banning the use of pack animals;
- raising the ticket price;
- devising a short supplementary trail.

Regular trail clean-ups are organised, but visitors complain of heavy litter and security problems, which impinge on the quality of their experience. A new government co-ordinating committee has been formed, but the Trail is still deteriorating very fast. The only solution seems to be to restrict visitor numbers by a quota booking system, but this is unlikely to be implemented since hiking the Inca Trail supports a huge range of businesses in Cuzco, as well as providing considerable local employment. At present, it looks as though further deterioration is inevitable, with trekkers moving elsewhere to less famous but better-managed walks.

REFERENCES

Ceballos-Lascurain, H. (1996) *Tourism, Ecotourism and Protected Areas.* Gland: IUCN.

Feilden, B.M. and Jokilehto, J. (1993) *Management Guidelines for World Cultural Heritage Sites.* Rome: ICCROM, UNESCO & ICOMOS.

Frost, P. and Bartle, J. (1995) *Machu Picchu Historical Sanctuary.* Lima: Neuvas Imagenas.

Hysop, J. (1990) *Inca Settlement Planning.* Austin: University of Texas Press.

Shackley, M. (1996) *Wildlife Tourism.* London: International Thomson Business Press.

Shackley, M. (1998) *Visitor Management: Case Studies from World Heritage Sites,* Oxford: Butterworth Heinemann.

von Droste, B., Plachter, H. and Rossler, M. (1995) *Cultural Landscapes of Universal Value.* New York: Fischer Verlag.

Dave McDonald
National Botanical Institute, Ecology and Conservation,
Private Bag X7, 7735 Claremont,
South Africa

Introduction: African mountains

Africa is the second largest continent in the world. The geology, climate, topography, biota and human inhabitants are diverse and, as a region and for its size, Africa is relatively undeveloped. Change is now occurring rapidly due to increasing human population pressures. The focus of this overview is the mountains of Africa, particularly their global context and how environmental and societal changes in Africa will affect them.

African mountains extend from the southern tip of the continent around the eastern edge of the continent (the 'Eastern Arc Mountains') and northwards into Ethiopia and the 'Horn'. Other mountains and highlands are found in Central and West Africa, north-western Africa and Africa north of the Sahara, the best-known of which are the Atlas Mountains. Despite the large extent of montane environments in Africa, investigations have been limited to those areas which are easily accessible. Most research has focused on the mountains of southern Africa, East Africa and Ethiopia, with less attention to the mountains of Central Africa and those in the north-west. A comprehensive overview of Africa and its centres of plant diversity is presented by Davis, Heywood and Hamilton (1994).

In African countries where mountains or highland plateaux are found, the highlands strongly influence the livelihoods of human populations. Highland areas, as well as being areas of high biodiversity, often have high agricultural potential and, as a result, high population densities. They can be free from disease (such as malaria) and provide opportunity for growing a range of cash and subsistence crops. Increasing population pressure and land fragmentation have resulted in deforestation, soil erosion and declining soil fertility in many areas. Issues that affect the lowlands affect the status of the mountains and vice versa. Therefore some of the key issues that need constant consideration in the context of the mountains are:

- conserving biodiversity with sustainable use of protected areas;
- understanding and building on local practices for improving soil, water and fertility management;
- integrating sustainable agricultural practices with conservation of biodiversity;
- ensuring political will to support programmes aimed at sustainable use of natural resources;

- the influence of climate change on ecosystems, particularly with respect to desertification;
- regional (cross-border) co-operation to foster conservation of mountain ecosystems on political boundaries.

Representation of African montologists at the Oxford conference was poor with only three papers being presented, two as posters and one in the Latin America / Africa workshop. Two of the papers focused on East Africa and covered very different subjects. Bakobi (in a paper not included in these proceedings) covered the biogeography of Tanzania with emphasis on the montane areas. As in many parts of Africa, the Tanzanian mountains are rich in species of plants and animals. These biota are being threatened by human activities, particularly agriculture, and statistics were provided for the rates of loss of natural habitats. Ellis-Jones and Tengberg described indigenous agricultural methods in Kenya, Tanzania and Uganda, and compared these with 'imported' methods. The problem of short-term higher productivity versus longer-term sustainable production was highlighted, and it was concluded that there is an urgent requirement for research with rural communities to build on their existing practices for soil and water conservation and fertility management. McDonald provided a general overview of the mountains of southern Africa, whose objective was to 'place southern African mountains on the map', since the global importance of these mountains is often not appreciated.

African montologists should be encouraged to participate more actively in the international arena for two reasons. Firstly, to expose the problems which African mountain ecosystems will face in the future so that these problems may be debated and discussed. In this process, an African perspective can also be brought into the global arena for more serious consideration. Secondly, African scientists need to learn from others with different experiences: such as those from developing countries such as India. In this way, the sustainable use and appreciation of mountains in Africa will be enhanced.

REFERENCE

S.D. Davis, V.H. Heywood and A.C. Hamilton (eds) (1994) *Centres of Plant Diversity: a Guide and Strategy for their Conservation.* Volume 1 Europe, Africa, South West Asia and the Middle East. Cambridge: IUCN.

Jim Ellis-Jones[1] and Anna Tengberg[2]
1 Silsoe Research Institute, Wrest Park
Silsoe, Bedfordshire, MK45 4HS, United Kingdom
2 Goteborg University, Department of Geosciences, Physical Geography
Guldhedsgatan 5A, 41381 Goteborg, Sweden

Modern methods from traditional conservation technologies: examples from hillside areas of Kenya, Tanzania and Uganda

Soil conservation has been strongly promoted in almost every developing country and over the last 50 years, and a large number of technically sound conservation technologies have been developed. As a consequence, many soil conservation projects and programmes have been implemented. Soil erosion has conventionally been perceived as the chief cause of land degradation, yet the limited effectiveness and low adoption of widely promoted anti-erosion measures make it necessary to reconsider the causes and therefore alternative measures to counter land degradation. Although soil erosion has been perceived as the main cause of land degradation, there has been an awareness that the underlying causes are social and economic in nature (Blaikie, 1985; Blaikie and Brookfield, 1987)

Farmers in many parts of Africa are using traditional or indigenous soil and water conserving (ISWC) practices as an integral part of their farming systems. Farmers have developed such methods which have maintained productivity and contributed to long-term sustainability, while introduced measures have often been rejected or simply failed to achieve their technical objectives.

Research in Kenya, Tanzania and Uganda has endeavoured to build on a new and an enlightened approach, based on the principle that development should be initiated from building on what farmers are already doing. Research has therefore utilised three basic approaches (Critchley et al., 1996):

- building on farmers' traditional practices;
- focusing on moisture conservation and fertility enhancement for plant growth, with conservation of soil being achieved simultaneously;
- involving farmers at all stages of the project from initial identification through planning, implementation, monitoring and evaluation through to dissemination of the results.

Different ISWC techniques have been studied in three areas, each with its own contrasting biophysical and socio-economic environments. In semi-arid Mbeere District, Kenya, the most striking characteristic is the environmental and socio-economic diversity, one of the underlying factors being the high rainfall

206

variability, leading to high variability in crop production. In the two mountainous areas of Mbinga, Tanzania and Kamwezi, Uganda, the climate is sub-humid and, although rainfall variability can be a problem, the priority problem in both areas is a long-term decline in soil productivity.

These varying characteristics result in differences in livelihoods and farming strategies between the areas. In Mbeere, risk management is the main priority with off-farm income opportunities increasingly seen as the best way to offset food and income shortages from farming. In contrast, cultivation of cash crops has become increasingly important in Mbinga and Kamwezi, where the maintenance of long-term soil productivity is the key natural resource problem. The ISWC practices studied included:

- Kenya: trash lines, stone bunds, log lines, Fanya juus and combinations of these techniques;
- Tanzania: ngoro pit and ridge systems with and without incorporation of organic matter;
- Uganda: trash lines, mulching and rainwater harvesting techniques.

Understanding the strengths and weaknesses of ISWC can assist in identifying which criteria determine farmers' use of different soil and water conserving practices. Most ISWC practices are characterised by their multiple functions, flexibility, spreading of labour requirements for construction and maintenance, and by gender differentiation of labour input (Reij *et al.*, 1996). The character-istics of ISWC and introduced technologies are compared in Table 1.

Table 1: Characteristics of ISWC and introduced technologies

Indigenous technologies	Introduced technologies
Integration of soil fertility enhancing, soil moisture and soil conserving techniques	Often specific to a single requirement
Integral component of land management and cropping systems	New management techniques may be required
Limited loss of land	Up to 20% of land area lost for cropping
Flexible	Inflexible
Low cost using existing materials and tools	Higher cost, additional tools
Provide an immediate payback	Often require an initial investment, annual maintenance with indeterminate benefits occurring over a long period
Can be labour intensive with gender specific activities, but flexibility allows use of labour at off-peak periods	Usually labour intensive with demand coinciding with existing labour peaks

An evaluation of the ISWC was carried out using both farmers' and researchers' criteria. Farmers' criteria are largely based on the quality of their natural resources (conserving those soils which give the highest return on investment), the resource level of their households (particularly income and labour availability), and cropping intensity, as well as cultural traditions related to age, education and gender. SWC which conserve moisture, reduce soil erosion, maintain soil fertility and, most importantly, increase productivity in the short term and are socially acceptable and economically viable, are those which farmers favour. ISWC practices often have these qualities, but farmers have not always been able to adjust their techniques to rapid social and economic changes affecting their farming systems and requiring increased intensity of land-use.

Soil productivity and economic modelling exercises have indicated that, although yield declines are lower when traditional technologies are used, long-term productivity remains a problem, especially for resource-poorest farmers. One of the main factors behind the decline in soil productivity seems to be decreasing fallow periods in combination with no or low external inputs. Nevertheless, there are encouraging numbers of farmers who have been able to maintain or increase the productivity of their lands. Therefore, an important way forward is to identify such farmer innovators, at all resource levels, who experiment within the framework of their existing farming systems using locally available materials. Such an approach to soil productivity enhancement is likely to build on the strengths of ISWC. Modern techniques need to encompass the flexibility of ISWC, providing options that can be modified and adopted to fit local biophysical and socio-economic circumstances. They will need to:

- enhance soil fertility and conserve soil and moisture;
- be an integral component of the farming system;
- use existing materials;
- not increase labour requirements at peak periods;
- address gender differences in labour utilisation;
- be low cost, not requiring a high initial investment and have an immediate impact on productivity;
- reduce risk.

REFERENCES

Blaikie, P. (1985) *The Political Economy of Soil Erosion in Developing Countries.* Longman, London.

Blaikie, P. and Brookfield, H. (1987) *Land Degradation and Society.* Methuen, London.

Critchley, W., Ellis-Jones, J., Kayombo, B., Kiome, R., Miiro, D. and Willcocks, T. (1996) Modern methods from traditional conservation techniques: An overview. Paper presented at 9th Conference of the International Soil Conservation Organization – Bonn, August 1996.

Reij, C., Scoones, I. and Toulmin, C. (1996) *Sustaining the Soil. Indigenous Soil and Water Conservation in Africa.* Earthscan Publications, London.

Dave McDonald
National Botanical Institute, Ecology and Conservation
Private Bag X7, 7735 Claremont
South Africa

The diversity of Southern African mountain ecosystems

The mountains of the Cape Region

Southern Africa's mountains form an arc extending from the Namibian border in the north-west, southwards to the Cape Peninsula, eastwards, inland from the coast, to the highest peaks in the Malotis and Drakensberg, and then north-eastwards to the Wolkberg. This arc of mountains forms the backbone of the Great Escarpment which bounds a high interior plateau. The mountains are a product of the break-up of Gondwana and subsequent significant erosion of the land surfaces (Partridge, 1997). The mountains in the north-west are in an arid zone with summer rainfall, whereas those in the south have a temperate climate with winter rainfall, and those in the east have a much more humid climate with summer rainfall.

Phytogeographically, southern Africa is divided into six regions (White, 1976; Goldblatt, 1978). The mountains in the southwest dominate the landscapes of the Cape Region. They are endowed with a rich flora, forming part of the sixth and smallest floral kingdom, the Cape Floral Kingdom. The vegetation of these mountains is mainly fynbos, a sclerophyllous, fire-prone shrubland characterised by graminoid *Restionaceae* and shrubby *Ericaceae* and *Proteaceae*. High levels of endemism are found in these shrublands: e.g., of the 1,228 species found in the flora of the southern Langeberg Mountains, a coastal range on the southern Cape coast, 150 species (13%) are endemic (McDonald and Cowling 1995). Afromontane forest is also found on these mountains, but is confined to moist habitats on south aspects.

The Cape mountains are highly dissected, with small rural communities and farms in the intervening valleys. The unpalatable mountain vegetation discourages the raising of livestock. Wild flowers are actively harvested for both the local and the export flower trade. Plants such as some species of *Restionaceae* are used for thatching, and some plant species are harvested for medicinal purposes. These activities are bound to increase as mountain areas become more accessible and human pressure increases. An additional threat to the natural veld is invasion by alien plants which outcompete the natural flora. They influence the vegetation by reducing species diversity and they significantly reduce water runoff from the mountain catchments.

The fynbos vegetation of the mountains does not support a diverse large mammal fauna, and the diversity of birds is also low. The herpetofauna is not diverse, but there are a number of endemic species. In contrast, there is a diverse insect fauna which displays a strong co-evolutionary affinity to the fynbos flora. Conservation of the Cape mountains with their diverse biota and complex habitats has presented an immense challenge to research. A concerted effort over the past 20 years has advanced our understanding of fynbos ecology, but many gaps in our knowledge still remain.

The KwaZulu-Natal Drakensberg and Malotis in Lesotho

The KwaZulu-Natal Drakensberg and Malotis (mountains) of Lesotho together form an elevated mountain massif, parts of which exceed 3000 m. They form the centre of the Afromontane Region in southern Africa with extensions to the Wolkberg in the north and the Amatola Mountains in the south. These mountains are in the summer rainfall zone. They experience heavy rains in the summer – often as dramatic thunderstorms – with contrasting dry, cold winters with moderate to heavy snowfalls.

At elevations exceeding 2,800 m, an alpine belt with dwarf ericaceous shrublands resembling those of fynbos of the Cape mountains is found. At lower elevations in the subalpine belt, the predominant vegetation is grassland including a mix of temperate C_3 and tropical C_4 grasses. Some woody communities are also found in this belt. Afromontane forest is confined to moist ravines and cool south-facing slopes where soils are better developed and where the forest communities are less exposed to fires.

It is estimated that there are 2,046 plant species in the KwaZulu-Natal Drakensberg (Killick 1994). Hilliard & Burtt (1987) recorded 1,375 species from the southern KwaZulu-Natal Drakensberg, 30% of which are endemic; from their study they concluded that emphasis should be placed on the importance of monocotyledons in this flora.

The KwaZulu-Natal Drakensberg and Malotis support a diverse array of fauna ranging from large antelope such as the Eland (*Taurotragus oryx*) to the diminutive dwarf chameleon (*Bradypodion dracomontanum*). Bird life is abundant and diverse: approximately 300 species have been recorded.

Lesotho, the 'Mountain Kingdom' is situated almost wholly in the mountains. Rural villages are scattered throughout the highlands and the inhabitants subsist on meagre agricultural crops of maize, wheat and peas. The wealth of the people is vested in their livestock; the subalpine and alpine vegetation is grazed by sheep, goats and cattle and evidence of overstocking is apparent in many places.

Of the Drakensberg Alpine Region (*sensu* Killick, 1994) 97% is conserved in reserves in the KwaZulu-Natal Drakensberg whereas only 3% is conserved in Lesotho. There is now a new initiative to create a "Trans-frontier Conservation Area" of 5,000 km^2 of the mountains which will straddle the north-eastern border between Lesotho and South Africa. This will greatly enhance the conservation of the alpine and subalpine ecosystems in these mountains.

REFERENCES

Goldblatt, P. (1978) An analysis of the flora of southern Africa: its characteristics, relationships and origins. *Annals of the Missouri Botanical Garden* **65**: 369–436.

Hilliard, O.M. and Burtt, B.L. (1987) The botany of the Southern Natal Drakensberg. *Annals of Kirstenbosch Botanic Gardens* **15**: 1–253.

Killick, D.J.B. (1994) Drakensberg Alpine Region: CPD Site AF82. Drakensberg Alpine Region, Lesotho and South Africa. In Davis, S.D., Heywood, V.H. and Hamilton, A.C. (eds) *Centres of Plant Diversity: a Guide and Strategy for their Conservation. Volume 1 Europe, Africa, South West Asia and the Middle East*, pp. 257–260. Cambridge: IUCN Publications Unit.

McDonald, D.J. and Cowling, R.M. (1995) Towards a profile of an endemic mountain fynbos flora: Implications for conservation. *Biological Conservation* **72**: 1–12.

Partridge, T.C. (1997) Evolution of landscapes. In Cowling, R.M., Richardson, D.M. and Pierce, S.M. (eds) *Vegetation of Southern Africa*. Cambridge University Press, Cambridge.

White, F. (1976) The vegetation map of Africa – the history of a completed project. *Boissiera* **24**: 659–666.

List of Acronyms

ALPNET – Network in Alpine Biodiversity (European Science Foundation)
ARTERI – Arctic-Alpine Terrestrial Ecosystems Research Initiative (European Commission DG XII)
CGIAR – Consultative Group on International Agriculture Research
CONDESAN – Consortium for the Sustainable Development of the Andean Ecoregion
EMF – European Mountain Forum
ENRICH – European Network for Research on Global Change (European Commission DG XII)
FAO – Food and Agriculture Organization of the United Nations
GCTE – Global Change and Terrestrial Ecosystems Core Project (IGBP)
GIS – Geographical Information System
ICIMOD – International Centre for Integrated Mountain Development
ICSU – International Council of Scientific Unions
IGBP – International Geosphere-Biosphere Programme (ICSU)
IGCP – International Geological Correlation Programme (UNESCO, IUGS)
IHDP – International Human Dimensions Programme (ISSC, ICSU)
INFOANDINA – Information System for the Andes (CONDESAN)
IPCC – Inter-governmental Panel on Climate Change
ISSC – International Social Science Council
IUBS – International Union of Biological Sciences
IUCN – World Conservation Union
IUGS – International Union of Geological Sciences
IUFRO – International Union of Forestry Research Organisations
IUMS – International Union of Microbiological Societies
MAB – Man and the Biosphere Programme (UNESCO)
MENRIS – Mountain Environment and Natural Resources Information System (ICIMOD)
ProClim – Forum for Climate and Global Change (Swiss Academy of Sciences)
SCOPE – Scientific Committee on Problems of the Environment (ICSU)
START – Global System for Research, Analysis and Training (IGBP, IHDP)
TSBF – Tropical Soil and Biological Fertility (IUBS, ICSU)
UNCED – United Nations Conference on Environment and Development
UNEP – United Nations Environment Programme
UNESCO – United Nations Educational, Scientific, and Cultural Organization

Index